MICROCOMPUTER APPLICATIONS IN OCCUPATIONAL HEALTH AND SAFETY

American Conference of Governmental Industrial Hygienists

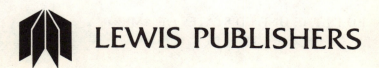

Library of Congress Cataloging-in-Publication Data

Microcomputer applications in occupational health and
    safety.

    Proceedings report the result of a symposium
sponsored by the American Conference of Governmental
Industrial Hygienists . . . in March 1986 — Pref.
    Bibliography: p.
    Includes index.
    1. Industrial hygiene — Data processing — Congresses.
2. Industrial safety — Data processing — Congresses.
3. Microcomputers — Congresses. I. American Conference
of Governmental Industrial Hygienists.
RC963.3.M53   1987   363.1′1′0285   86-27676
ISBN 0-87371-078-9

COPYRIGHT © 1987 by LEWIS PUBLISHERS, INC.
ALL RIGHTS RESERVED

Neither this book nor any part may be reproduced or transmitted in
any form or by any means, electronic or mechanical, including
photocopying, microfilming, and recording, or by any information
storage and retrieval system, without permission in writing from the
publisher.

LEWIS PUBLISHERS, INC.
121 South Main Street, P.O. Drawer 519, Chelsea, Michigan 48118

PRINTED IN THE UNITED STATES OF AMERICA

PREFACE

Industrial hygienists solve problems through the recognition, evaluation, and control of work-related illness. As integrators and communicators of the complex information that this mission demands, industrial hygienists are vigilant in finding new tools to help them. It is not surprising, then, to find industrial hygienists adopting microcomputers, one of technology's most recent and exciting offerings, to aid them in these tasks.

This book reports the results of a symposium sponsored by the American Conference of Governmental Industrial Hygienists (ACGIH) in March 1986. The symposium provided a forum for industrial hygiene microcomputer users, at all experience levels, to share and to learn how others use this new and powerful technological tool. State-of-the-art applications presented by experienced industrial hygienists from government, industry, and unions featured practical, "hands-on," microcomputer applications. Plenary and poster session topics ranged from demonstrations of how to use a simple database for tracking industrial hygiene personal samples, to electronic spreadsheets used for industrial hygiene laboratory control, to modeling for indoor air pollution, to programs for ergonomic design.

The innovative uses of microcomputer presented here are precursors of advances in microcomputing technology, both software and hardware, that will come with industrial hygienists' use of them. This book should help guide industrial hygienists toward this future.

CONTENTS

SECTION I
RECENT TRENDS

1. Microcomputers: A Powerful Tool for Industrial Hygienists and Safety Professionals, *Brad T. Garber* 1
2. Information System Building Blocks, *Charles W. McNichols* 27
3. Systems Implementation, *Howard L. Kusnetz* 59
4. Dependable Information Can Define the Environment, *Rafael H. Estupinian* 69
5. Automation of Materials Inventory and MSDS Information, *Andrew J. Becker* 79

SECTION II
NEW PRODUCTS

6. The Evolution of Occupational Health Information Systems—An Historical Perspective, *Wanda Rappaport* 91
7. Health Surveillance: Managing "Known" and "Unknown" Risk Through Integration of Industrial Hygiene and Health Data, *Richard Besserman* 103

SECTION III
PRACTICAL USES

8. Electronic Spreadsheets in Industrial Hygiene Management, *Mitchell S. Bergner* 113
9. Data Base Management, *Denese A. Deeds* 121
10. Word Processing Applications in Industrial Hygiene, *Leonard Wilcox* 127
11. Computer Communications, *Charles M. Baldeck* 133

SECTION IV
EXAMPLE APPLICATIONS

12. A Lotus 1-2-3 Data Base Model for Keeping and Tracking Citations, *Bryan A. Hardin* 155
13. Computer Aided Design (CAD) Applications in Industrial Hygiene, *Doan J. Hansen* 159
14. Researching Toxic Substances Information Through the Use of Interrelational Data Base, *Mitchell Brathwaite* 165
15. Metabolic Energy Expenditure Prediction and Static Strength Prediction Program Models, *Charles Wooley and Randy Raybourn* 169
16. Example Problems—Computer-Aided Ventilation Design, *David E. Clapp* 179
17. Use of a Microcomputer Spreadsheet Program to Model Carbon Monoxide Levels and Effects Due to Indoor Construction with Partial General Ventilation, *L. W. Whitehead* 193
18. The Safety and Health Aspects of Video Display Terminals, *Scott E. Merkle* 197
19. A Computerized System for Tracking Field Equipment Maintenance, *Martin T. Abell* 201
20. Computer Applications Glossary, *Kenneth S. Cohen* 209

Index 223

SECTION I

Recent Trends

CHAPTER 1

MICROCOMPUTERS: A POWERFUL TOOL FOR INDUSTRIAL HYGIENISTS AND SAFETY PROFESSIONALS

BRAD T. GARBER, PhD

Professor, Occupational Safety and Health, University of New Haven, West Haven, Connecticut

INTRODUCTION

Microcomputers are powerful, cost-effective tools for performing many occupational safety and health tasks. Some of these tasks are "naturals" for which the microcomputer can be applied easily and efficiently (Table I). Others require a significant amount of thought and planning or the results are likely to be unsatisfactory (Table II).

DATA BASE MANAGEMENT

Data bases are collections of records that have a common format. Programs which are used to efficiently process large numbers of records are called data base management programs. The functions they perform are (1) creation of new data bases, (2) data entry, (3) data modification, and (4) data retrieval (searching, sorting, report generation).

Data base management applications fall into the category of tasks which require a great deal of diligence and effort to obtain satisfactory results. Planning should include careful consideration of the following

TABLE I. OSH Tasks which can be Performed Easily and Effectively on a Microcomputer

Good Applications

Safety statistics	Ventilation calculations
Air sampling calculations	Biological monitoring data
Noise calculations	Accessing mainframe computers
Word processing	Budgets

TABLE II. OSH Tasks which Require Significant Planning

Good Applications but Proceed with Caution

Air sampling data management
Medical data management
Comprehensive exposure/medical systems

factors: record structure, data quality, expense of data entry, hardware needs, and software needs.

Creation of Data Bases

In order for a computer to efficiently handle large volumes of information, data should be structured. It is crucial that the record format to be used be fully defined and well thought out before data entry is performed. It is often difficult and tedious to make changes after a data base has been created and a large amount of data entered. This is especially true if the changes involve adding new information to records.

The importance of record structure is strongly reflected in data entry, which is frequently a time-consuming and therefore expensive task. Record structure should be designed to facilitate data entry. One way to accomplish this is to eliminate any extraneous or redundant information. Another is to use coding schemes. Coding is the process of assigning short combinations of characters to represent words or phrases. Coding not only speeds entry but also decreases the amount of disk space required to store a record. Where data is coded it is useful to have a dictionary function which retranslates the codes. This is especially desirable in report generation.

The individual pieces of information stored in records are called fields. Data base management programs usually require the user to specify the length of each field and its data type (Table III). The two most common data types are numeric (numbers) and alphanumeric (combinations of numbers and letters). Specifying the data type allows the computer to more efficiently store and manipulate data.

Data Entry

Good data entry procedures are crucial to the successful functioning of a data base management system. They should be designed to maximize speed of entry and to minimize the likelihood of error. The procedures should also minimize stress to the data entry operator.

One technique which is useful in attaining these objectives is to have the equivalent of a blank data entry form appear on the computer's video display screen. After the operator enters data for a specific field he moves to the next field with a single keystroke (Figure 1).

It is advantageous to have the computer check for obvious errors. Sophisticated error checking schemes can reduce the number of errors significantly. Typical of the checks which can be performed are:

1. Making sure numbers are entered where numbers are expected and letters are entered where letters are expected.
2. Determining if numerical values are within a specific range.
3. Determining if alphabetical entries have specified acceptable values.

The quality and quantity of checking that is performed is an important determinant of how error free the data will be.

Data Modification

It is sometimes necessary to change records. For individual records the procedure is very straightforward. The record is read from the disk, modified, and written back out to disk. When errors exist in many records, fixing them one at a time can be very cumbersome. As a result most good data base management programs contain routines which automatically fix all records which contain a specific systematic error.

Searching

The primary advantage of using a computer for data base management is its ability to quickly search through large numbers of records and cull out those which meet user defined criteria. When just one criterion is

TABLE III. Example of a Record Structure for a Company Telephone Book

Field 1 Last Name (Alphanumeric)	Field 2 First Name (Alphanumeric)	Field 3 Department (Alphanumeric)	Field 4 Extension (Numeric)
J O N E S	J O H N	S A F E T Y	3 2 6
1 2 3 4 5	10 11 12 13	16 17 18 19 20 21 22 23 24 25 26 27 28	29 30 31 32

Character Number: 1 2 3 4 5   6 7 8 9   10 11 12 13 14 15   16 17 18 19 20 21   22 23 24 25 26 27 28   29 30 31 32

Record structure: Field 1 = 15 characters
 Field 2 = 8 characters
 Field 3 = 6 characters
 Field 4 = 3 characters
 Total record length = 32 characters

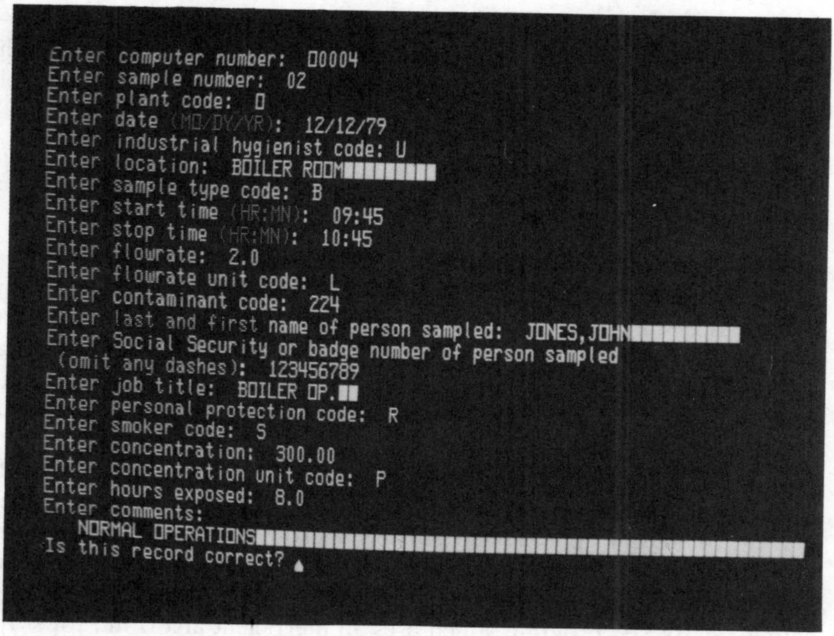

```
Enter computer number: 00004
Enter sample number: 02
Enter plant code: D
Enter date (MO/DY/YR): 12/12/79
Enter industrial hygienist code: U
Enter location: BOILER ROOM
Enter sample type code: B
Enter start time (HR:MN): 09:45
Enter stop time (HR:MN): 10:45
Enter flowrate: 2.0
Enter flowrate unit code: L
Enter contaminant code: 224
Enter last and first name of person sampled: JONES,JOHN
Enter Social Security or badge number of person sampled
   (omit any dashes): 123456789
Enter job title: BOILER OP.
Enter personal protection code: R
Enter smoker code: S
Enter concentration: 300.00
Enter concentration unit code: P
Enter hours exposed: 8.0
Enter comments:
   NORMAL OPERATIONS
Is this record correct?
```

Figure 1. Photograph of screen input format.

used the procedure is quite easy. For example, in an industrial hygiene air sampling data base one might want to search for all lead samples taken. The process is quite a bit more complicated when several criteria or complex conditions are specified. For instance, one might want to search for all lead samples with concentrations greater than 0.05 mg/m^3 taken in 1982. Such queries must be defined very carefully or errors are likely to occur. Sophisticated data base management programs have their own query languages for defining searches. The rules which define the acceptable syntax must be adhered to carefully or the program may abort or, still worse, print out records that meet criteria other than those desired.

Deleting Records

Many data base management programs delete records by tagging them as invalid rather than by actually removing them from the disk. This serves at least two purposes. First, it increases the speed of record deletion. Second, the deleted records can be examined and restored if necessary. Because this procedure is wasteful of disk space, most programs will allow the user to periodically run a special routine which actually removes invalid records and thus frees up space for additional records.

Sorting

Sorting refers to putting data into a predefined order. Usually this means alphabetizing alphabetic data or putting numeric data in ascending or descending order. A desirable feature of a data base management program is the ability to sort on more than one field. For instance, if a sort of the primary field produces several large groupings of records that have the same field value, the user may want to further sort the records within each group using a second field (Tables IV, V, VI).

Sorting can be a very time-consuming process for a computer. If much sorting is to be done, it is essential that an efficient algorithm (a mathematical or logical approach to solving a problem) be used. It is also important that a language which produces fast running programs be used to write the sort routine. For this reason, assembly language is usually chosen. A BASIC program which uses an inefficient algorithm for sort-

TABLE IV. Unsorted Data

Plant	Location	Air Concentration (mg/m^3)
Boston	Grinding Room	3.1
Denver	Grinding Room	4.2
Boston	Bagging Room	3.7
Boston	Warehouse	1.1
Denver	Bagging Room	2.1
Houston	Grinding Room	3.7
Philadelphia	Grinding Room	5.6
Boston	Loading Dock	1.1
Denver	Warehouse	1.5
Denver	Loading Dock	1.5
Philadelphia	Bagging Room	2.7
Philadelphia	Warehouse	1.6
Boston	Office	0.9
Denver	Office	0.1
Houston	Bagging Room	2.2
Philadelphia	Loading Dock	1.7
Houston	Warehouse	1.9
Philadelphia	Office	0.3
Houston	Loading Dock	1.3
Houston	Office	0.4

TABLE V. Average Air Concentrations of Total Dust Sorted by Concentration

Plant	Location	Air Concentration (mg/m³)
Philadelphia	Grinding Room	5.6
Denver	Grinding Room	4.2
Boston	Bagging Room	3.7
Houston	Grinding Room	3.7
Boston	Grinding Room	3.1
Philadelphia	Bagging Room	2.7
Houston	Bagging Room	2.2
Denver	Bagging Room	2.1
Houston	Warehouse	1.9
Philadelphia	Loading Dock	1.7
Philadelphia	Warehouse	1.6
Denver	Warehouse	1.5
Denver	Loading Dock	1.5
Houston	Loading Dock	1.3
Boston	Warehouse	1.1
Boston	Loading Dock	1.1
Boston	Office	0.9
Houston	Office	0.4
Philadelphia	Office	0.3
Denver	Office	0.1

TABLE VI. Average Air Concentrations of Total Dust Sorted by Location as Primary Key and Concentration as Secondary Key

Plant	Location	Air Concentration (mg/m³)
Boston	Bagging Room	3.7
Boston	Grinding Room	3.1
Boston	Warehouse	1.1
Boston	Loading Dock	1.1
Boston	Office	0.9
Denver	Grinding Room	4.2
Denver	Bagging Room	2.1
Denver	Warehouse	1.5
Denver	Loading Dock	1.5
Denver	Office	0.1
Houston	Grinding Room	3.7
Houston	Bagging Room	2.2
Houston	Warehouse	1.9
Houston	Loading Dock	1.3
Houston	Office	0.4
Philadelphia	Grinding Room	5.6
Philadelphia	Bagging Room	2.7
Philadelphia	Loading Dock	1.7
Philadelphia	Warehouse	1.6
Philadelphia	Office	0.3

ing can take many hours to perform a sort that would be accomplished in a few minutes by an assembly language program which uses an efficient algorithm.

PROTECTION OF DATA

Much time and expense goes into data entry. Inadvertent destruction of that data could therefore be very costly. There are several ways in which data could be lost. These include (1) physical damage to the storage media, (2) exposure of magnetic media to magnetic fields, (3) errors in a program, (4) power loss or fluctuations during critical disk operations, (5) accidental activation of an erase function, (6) physical damage to the computer, (7) static electricity, and (8) accidental contamination of a data base with bad data.

One method of protecting against the catastrophic loss of data is to make backup copies frequently. For critical data several copies should be made and at least one stored off-site in a secure place which is free from magnetic fields or other sources of damage. The size of the data base and the importance of the information stored will dictate how extensive the backup procedures should be. The formulation of adequate backup procedures is among the most important aspects of properly using a computer system because it is so very easy to destroy many hours of work in a few seconds.

Static electricity can be a major problem for computer users. Not only can it cause the loss of data on magnetic media, but it can also destroy electronic parts. Although the currents generated when static electricity is discharged are quite small, the voltages are very high. As a result, sensitive electronic components can be damaged. The computer environment should be kept as free from static electricity as possible. Some of the means of accomplishing this are keeping the air from becoming too dry (by using humidifiers), placing the operator's chair on a static draining mat and using static dissipating sprays on rugs.

Changes in the characteristics of the power source supplying a computer can cause data loss, malfunction, or in extreme cases, actual physical damage. Power surges and spikes are the two most common problems. These are often caused by the intermittent operation of heavy equipment. A number of devices which condition the source of power are available for dealing with these problems. They are not always necessary, but where needed, can make the difference between a smoothly operating system and one subject to periodic data loss and component failure.

Airborne contaminants can interfere with disk drive operation. Corrosive gases or particulate matter are especially troublesome. High concentrations of cigarette smoke can be a significant source of faulty disk operation.

Confidentiality of Data

Industrial hygiene data bases can contain information considered to be confidential. For example, biological monitoring data are medical in nature and should be treated as confidential medical records.

There are several ways to protect against unauthorized access to confidential data. One is to require a secret password be entered before a data base can be accessed. This method is widely used on mainframes and minicomputers but has not been widely adopted for microcomputers. Disadvantages of this method are the following:

1. Passwords are difficult to protect. Someone simply glancing over a computer user's shoulder can easily steal his password.
2. The supervisor who controls operation of the computer usually has access to all data files and passwords.
3. A particularly clever programmer may be able to figure out a way to bypass the password protection.

A second means of ensuring confidentiality is to store the data base on removable media. The media (tape, disk, etc.) can be taken to a site remote from the computer and stored in a secure place such as a safe. This method is especially useful for microcomputers since the most common storage medium is the floppy disk which is easily transported and takes up little space. A third way to ensure confidentiality is to encrypt the data. The decode scheme should be accessible only to the authorized user and should be stored on removable media.

In general, it is much easier to maintain confidentiality when using a microcomputer than when using a mainframe or mini. With a microcomputer the user can fully control the operating environment. One does not have to contend with operators, other users, supervisors, and data-processing professionals, all of whom have the potential to access confidential data. With a microcomputer, the data in its solid state memory can be erased quickly and easily by simply pulling the plug. Data on floppies can be protected by storing them in a place accessible only to legitimate users.

EXAMPLE OF HOW TO SET UP AND USE AN INDUSTRIAL HYGIENE DATA BASE

The most practical way to set up an industrial hygiene data base on a microcomputer is through the use of one of the many good data base management program packages which are commercially available. One such program, DBASEII, can be used for the management of air sampling data. While it is beyond the scope of this chapter to provide a comprehensive tutorial on DBASEII, the material presented will exhibit many of its features, thereby giving the reader an overview of how a data base management program works.

The data structure that will be used is the following:

Field	Characters
Computer number	1–5
Sample number	6–7
Plant code	8
Date of sampling	9–14
Industrial hygienist code	15
Location of sampling	16–35
Sample type code	36
Sample start time	37–40
Sample stop time	41–44
Flowrate	45–47
Flowrate unit	48
Contaminant code	49–51
Person sampled (name)	52–71
Social security number	72–80
Job title	81–92
Personal protection code	93
Smoker code	94
Concentration	95–102
Concentration unit code	103
Hours exposed	104–106
Comments	107–176

Total record contains 176 characters.

Each of the fields will be defined as alphanumeric, except sample number and concentration which will be numeric (strictly numbers). Alphanumeric fields are called character fields by DBASEII because they can consist of alphabetic or numeric characters. The symbol for character fields is "C" and for numeric fields is "N". The data base is created as follows:

```
. CREATE IHDATA
ENTER RECORD STRUCTURE AS FOLLOWS:
FIELD    NAME,TYPE,WIDTH,DECIMAL PLACES
001      COMPNUM,C,5
002      SAMPNUM,N,2
003      PLANTCODE,C,1
004      SAMPDATE,C,6
005      INDHYGCODE,C,1
006      SAMPLOC,C,20
007      SAMPTYPE,C,1
008      SAMPSTART,C,4
009      SAMPSTOP,C,4
010      FLOWRATE,C,3
011      FLOWUNIT,C,1
012      CONTAMCODE,C,3
013      PERSONSAMP,C,20
014      SSNUM,C,9
015      JOBTITLE,C,12
016      PERSPROT,C,1
017      SMOKERCODE,C,1
018      CONCENTRAT,N,8,3
019      CONCUNIT,C,1
020      HOURSEXP,C,3
021      COMMENTS,C,70
022
INPUT DATA NOW? N
```

The command CREATE IHDATA followed by a carriage return caused the computer to request that the user enter the data structure. For those fields defined as numeric, it is necessary to specify the number of decimal places. If no entry is made the default value is zero. When the computer requested information for field 22 and the carriage return key was depressed without entering any data, the computer recognized that field 21 was the last one in the record. With DBASEII most data entry operations and commands must be followed by a carriage return.

To verify that the structure was entered correctly, the command DISPLAY STRUCTURE can be used:

```
. DISPLAY STRUCTURE
NO DATABASE FILE IN USE, ENTER FILENAME: IHDATA
STRUCTURE FOR FILE:           C:IHDATA   .DBF
NUMBER OF RECORDS: 00000
DATE OF LAST UPDATE: 02/14/84
PRIMARY USE DATABASE
FLD        NAME              TYPE    WIDTH      DEC
001        COMPNUM            C       005
002        SAMPNUM            N       002
003        PLANTCODE          C       001
```

004	SAMPDATE	C	006	
005	INDHYGCODE	C	001	
006	SAMPLOC	C	020	
007	SAMPTYPE	C	001	
008	SAMPSTART	C	004	
009	SAMPSTOP	C	004	
010	FLOWRATE	C	003	
011	FLOWUNIT	C	001	
012	CONTAMCODE	C	003	
013	PERSONSAMP	C	020	
014	SSNUM	C	009	
015	JOBTITLE	C	012	
016	PERSPROT	C	001	
017	SMOKERCODE	C	001	
018	CONCENTRAT	N	008	003
019	CONCUNIT	C	001	
020	HOURSEXP	C	003	
021	COMMENTS	C	070	
** TOTAL **			00177	

Note that the total record length will be 177 bytes which is one greater than the number of characters in the fields. Records always contain one byte (character) of overhead.

To access a specific data base the command USE is employed. The following brings up the IHDATA data base:

Records are changed by using the EDIT command. The cursor (symbol on the screen which shows where letters typed on the keyboard will appear) is moved to the desired place and the change is made:

. EDIT RECORD 1

```
RECORD # 00001
RECORD # 00001
COMPNUM    :O0001:
SAMPNUM    :03:
PLANTCODE  :O:
SAMPDATE   :112978:
INDHYGCODE :A:
SAMPLOC    :CRANE #1            :
SAMPTYPE   :C:
SAMPSTART  :0750:
SAMPSTOP   :1512:
FLOWRATE   :1.4:
FLOWUNIT   :L:
CONTAMCODE:001:
PERSONSAMP:JONES,JOHN           :
SSNUM      :123456789:
JOBTITLE   :CRANE OP    :
```

```
PERSPROT    :*:
SMOKERCODE:S:
CONCENTRAT :0.052:
CONCUNIT    :M:
HOURSEXP    :***:
COMMENTS    :
            :
```

The EDIT function is exited by depressing the control and W keys simultaneously to record the changes, or depressing the control and Q keys to ignore the changes.

Records are added by using the APPEND command:

```
. APPEND

RECORD # 00102

RECORD # 00102
COMPNUM     :N0012:
SAMPNUM     :12:
PLANTCODE   :N:
SAMPDATE    :121283:
INDHYGCODE :F:
SAMPLOC     :BOILER ROOM          :
SAMPTYPE    :O:
SAMPSTART   :0815:
SAMPSTOP    :1545:
FLOWRATE    :2.0:
FLOWUNIT    :L:
CONTAMCODE:002:
PERSONSAMP:JONES,JOHN            :
SSNUM       :123456789:
JOBTITLE    :BOILER OP            :
PERSPROT    :R:
SMOKERCODE:S:
CONCENTRAT :300.000:
CONCUNIT    :P:
HOURSEXP    :8.0:
COMMENTS    :NORMAL OPERATIONS
            :
```

Records can be viewed by using either the DISPLAY or the LIST command. An entire record can be seen as follows:

```
. LIST RECORD 1
00001   00001   3  O  112978  A  CRANE #1    C  0750 1512 1.4 L 001 JONES,JOHN
              123456789 CRANE OP         * S 0.052 M ***
```

If just certain fields of the records are desired, a command such as the following can be issued:

```
. LIST NEXT 10 PERSONSAMP SSNUM JOBTITLE CONCENTRAT CONTAMCODE
00001   JONES,JOHN      123456789  CRANE OP      0.052  001
00002   JONES,JOHN      123456789  CRANE OP      0.018  002
00003   JONES,JOHN      123456789  CRANE OP      4.595  003
00004   JONES,JOHN      123456789  CRANE OP      0.050  001
00005   JONES,JOHN      123456789  CRANE OP      0.016  002
00006   JONES,JOHN      123456789  CRANE OP      1.983  003
00007   JONES,JOHN      123456789  CRANE OP      0.033  001
00008   JONES,JOHN      123456789  CRANE OP      0.020  002
00009   JONES,JOHN      123456789  CRANE OP      2.069  003
00010   JONES,JOHN      123456789  CRANE OP      0.026  001
```

Records that meet criteria chosen by the user can be selected. For instance, if one wanted information on samples taken for a specific contaminant, (in this case the one with code 002), the following command would be issued. The desired information is computer number, name of person sampled, job title and concentration:

```
. LIST ALL COMPNUM PERSONSAMP JOBTITLE CONCENTRAT FOR
CONTAMCODE = '002'
00002  O0002  JONES,JOHN       CRANE OP     0.018
00005  O0005  JONES,JOHN       CRANE OP     0.016
00008  O0008  JONES,JOHN       CRANE OP     0.020
00011  O0011  SMITH,ROBERT     CRANE OP     0.099
00014  O0014  SMITH,ROBERT     CRANE OP     0.011
00017  O0017  SMITH,ROBERT     CRANE OP     0.071
00020  O0020  ADAMS,THOMAS     CRANE OP     0.325
00023  O0023  ADAMS,THOMAS     CRANE OP     0.114
00026  O0026  ADAMS,THOMAS     CRANE OP     0.209
00029  O0029  ADAMS,THOMAS     CRANE OP     0.192
00032  O0032  ADAMS,THOMAS     CRANE OP     0.435
00034  O0034  ADAMS,THOMAS     CRANE OP     0.001
00038  O0038  ADAMS,THOMAS     CRANE OP     0.505
00043  O0043  ADAMS,THOMAS     CRANE OP     0.007
00046  O0046  DOE,PAUL         CRANE ASST   0.014
00049  O0049  DOE,PAUL         CRANE ASST   0.018
00052  O0052  DOE,PAUL         CRANE ASST   0.032
```

If the user knows the contents of just a portion of the field which is to be the subject of a search he still can locate the desired records. For example, if all of the records for someone with the last name Jones are wanted but the first name is not known, we can search on the PERSON-

SAMP field using the selection criteria that the word "JONES" be contained anywhere within the field:

```
. LIST PERSONSAMP CONTAMCODE CONCENTRAT CONCUNIT FOR
'JONES'$PERSONSAMP
00001   JONES,JOHN              001     0.052   M
00002   JONES,JOHN              002     0.018   M
00003   JONES,JOHN              003     4.595   M
00004   JONES,JOHN              001     0.050   M
00005   JONES,JOHN              002     0.016   M
00006   JONES,JOHN              003     1.983   M
00007   JONES,JOHN              001     0.033   M
00008   JONES,JOHN              002     0.020   M
00009   JONES,JOHN              003     2.069   M
00010   JONES,JOHN              001     0.026   M
```

'JONES'$PERSONSAMP means JONES within the field Personsamp. It is possible to search using several criteria:

```
. LIST FOR PERSONSAMP = 'JONES,JOHN'.AND.CONTAMCODE = '002'
PERSONSAMP CONTAMCODE CONCENTRAT
00002   JONES,JOHN      002     0.018
00005   JONES,JOHN      002     0.016
00008   JONES,JOHN      002     0.020
```

.AND. is referred to as a Boolean (logical) operator. Other Boolean operators are ".NOT." and ".OR."

A file can be sorted. For instance, if it is desired to group all of the samples for each contaminant code, a sort on the contaminant code field could be performed:

```
. SORT.ON CONTAMCODE TO SORTFILE
SORT COMPLETE
. USE SORTFILE
. LIST ALL CONTAMCODE CONCENTRAT
00001   001     0.285
00002   001     0.009
00003   001     0.220
00004   001     0.533
00005   001     0.275
00006   001     0.505
00007   002     300.000
00008   003     4.644
00009   003     0.032
00010   003     0.410
00011   003     0.607
00012   003     0.719
00013   003     0.844
```

```
00014    003    1.168
00015    004    0.239
00016    004    0.387
00017    004    0.048
00018    004    0.273
00019    004    0.118
00020    004    0.187
00021    004    0.478
```

The BROWSE command allows the user to view the entries in the data base. The output is formatted in an easy to read fashion. Data can be modified by moving the cursor to the proper position on the screen and typing over characters which are to be changed. The changes are made permanent by depressing the control and W keys simultaneously:

```
. BROWSE

RECORD # :00001
COMPN   SA   O   SAMPDA   I   SAMPLOC--------   S   SAMP   SAMP   FLO   F   CON
O0001    3   O   112978   A   CRANE #1          C   0750   1512   1.4   L   001
O0004    4   O   112978   A   CRANE #1          C   0753   1513   1.6   L   001
O0007    5   O   112978   A   CRANE #1          C   0757   1512   1.3   L   001
O0010    6   O   112978   A   CRANE #1          C   0928   0938   1.6   L   001
O0013    7   O   112978   A   CRANE #2          C   1326   1334   1.6   L   001
O0016    8   O   112978   A   CRANE #2          C   0948   0953   1.6   L   001
O0019    9   O   112978   A   CRANE #2          C   0940   0947   1.6   L   001
O0022   10   O   112978   A   CRANE #3          C   1031   1048   1.6   L   001
O0025   11   O   112978   A   CRANE #3          C   0948   0954   1.8   L   001
O0028   12   O   112978   A   CRANE #3          C   1100   1116   1.6   L   001
O0031   13   O   112978   A   CRANE #3          C   1432   1444   1.6   L   001
O0035   14   O   112978   A   CRANE #3          2   0811   1514   9.7   F   001
O0039   15   O   112978   A   CRANE #3          2   0813   1514   3.9   F   001
O0042    1   O   113078   A   CRANE #3          C   0740   1545   2.1   L   001
O0045    2   O   113078   A   CRANE #1          C   0750   1455   1.3   L   001
O0048    3   O   113078   A   CRANE #1          C   0752   1453   1.1   L   001
O0051    4   O   113078   A   CRANE #1          C   0754   1455   1.6   L   001
O0054    5   O   113078   A   CRANE #1          C   0815   1453   1.4   L   001
O0057    6   O   113078   A   CRANE #2          C   0937   0955   1.8   L   001
```

As records are added they are assigned a sequential record number indicating their position in the data base. For example, the first 10 records entered would be assigned the number 1 to 10. The response to many of the commands, such as DISPLAY NEXT n which displays the next n records, depends on the user's current position within the data base. Position can be changed using the GO, GOTO, and SKIP commands.

— GO TOP positions at the beginning of the data base
— GOTO n positions at record n
— simply typing a number positions at that record number
— SKIP n moves the position down n records.

Examples are:

	Position in Data Base
. GO TOP	record 1
. GOTO 6	record 6
. GO TOP	
. SKIP 3	
RECORD: 00004	record 4
. 10	record 10

DBASEII can be used to perform calculations by using the "?" command. For example:

. ? 200/3
  66

Results of calculations can be stored to variables as follows:

. STORE 100/2 TO S
  50

S now has the value 50.

Records are removed using the DELETE command:

. DELETE RECORD 10
00001 DELETION (S)

deletes record 10.

. DELETE ALL
00093 DELETION(S)

deletes all of the records.

. DELETE NEXT 10
00010 DELETION(S)

deletes the next 10 records.

DELETE does not actually remove the records from the disk but simply flags (labels) them as deleted. They can be reinstated using the RECALL command:

```
. RECALL ALL
00103 RECALL(S)
```

recalls all deleted records.

The number of records in a data base can be determined using the COUNT command:

```
. COUNT
COUNT = 00102
```

Normally when records are added they are positioned directly after the last record in the existing data base, and are therefore assigned the record number n + 1 where n is the current number of records. To position the record elsewhere, the INSERT command is used. The following sequence of commands would insert a blank record between existing records 10 and 11:

```
. GOTO 10
. INSERT BLANK
```

Alternately:

```
. 10
. INSERT BLANK
```

would accomplish the same thing.

Disk space can be freed up for use by other records by using the PACK command after a DELETE. PACK makes the records unavailable for future use.

The structure of a data base can be changed by using the MODIFY command as follows:

```
. MODIFY STRUCTURE
MODIFY ERASES ALL DATA RECORDS ... PROCEED? (Y/N) Y
```

	NAME	TYP	LEN	DEC	
FIELD 01	:COMPNUM	C	005	000	:
FIELD 02	:SAMPNUM	N	002	000	:
FIELD 03	:PLANTCODE	C	001	000	:
FIELD 04	:SAMPDATE	C	006	000	:
FIELD 05	:INDHYGCODE	C	001	000	:
FIELD 06	:SAMPLOC	C	020	000	:
FIELD 07	:SAMPTYPE	C	001	000	:
FIELD 08	:SAMPSTART	C	004	000	:
FIELD 09	:SAMPSTOP	C	004	000	:
FIELD 10	:FLOWRATE	C	003	000	:
FIELD 11	:FLOWUNIT	C	001	000	:

```
FIELD 12   :CONTAMCODE    C   003   000   :
FIELD 13   :PERSONSAMP    C   020   000   :
FIELD 14   :SSNUM         C   009   000   :
FIELD 15   :JOBTITLE      C   012   000   :
FIELD 16   :PERSPROT      C   001   000   :
FIELD 17   :SMOKERCODE    C   001   000   :
FIELD 18   :CONCENTRAT    N   008   003   :
FIELD 19   :CONCUNIT      C   001   000   :
FIELD 20   :HOURSEXP      C   003   000   :
FIELD 21   :COMMENTS      C   070   000   :
FIELD 22   :
```

Changes are made by typing over the old entry. When all of the desired changes have been entered, they can be permanently recorded by depressing the control and W keys simultaneously.

Data files can be duplicated using the COPY command:

```
. USE IHDATA
. COPY TO IHDATA2
00100 RECORDS COPIED
00102 RECORDS COPIED
```

In the above example, a duplicate of the data base IHDATA called IHDATA2 was produced.

When it is necessary to systematically change a group of records, the REPLACE command can be used:

```
. REPLACE ALL PLANTCODE WITH 'A' FOR PLANTCODE = 'O'
00100 REPLACEMENT(S)
```

The above would change the plant code to "A" in all records with plant code "O".

The CHANGE command can be used to modify specific fields in a group of records:

```
. CHANGE NEXT 2 FIELD PLANTCODE

RECORD: 00001

PLANTCODE: A
CHANGE? A
TO       H

PLANTCODE: H
CHANGE?
```

RECORD: 00002

PLANTCODE: H
CHANGE? H
TO   A

PLANTCODE: A
CHANGE?

    The above command changed the plant code from A to H in the first 2 records.
    Searching through a data base for records that meet user defined criteria normally requires that every record be read. By indexing the data base on those fields which are frequently the subject of a search, locating records can be speeded. For example, if it is frequently necessary to cull out the records for particular plants the data base could be indexed on plant code. If the user wanted to index on the name of the person sampled it could be done as follows:

```
. INDEX ON PERSONSAMP TO PS
00100 RECORDS INDEXED
00102 RECORD INDEXED
```

    When a field has been indexed it is possible to use the FIND command which locates the first record whose indexed field contains a series of user selected characters:

```
. USE IHDATA INDEX PS
. FIND JONES
. DISPLAY
00001    O0080     13 A 113078 A CRANE #3    C 1037 1056 1.6 L 003 JAMES,JOHN
                   678901234 CRANE ASST      * S      4.644 M ***
```

    In the above example the data base was indexed on the personsamp field. The FIND JONES command showed the first record which contained the word JONES anywhere in the personsamp field. In this instance it was record 1.
    A useful feature of DBASEII is the ability to generate and store report formats. The REPORT command performs this function:

```
. REPORT
ENTER REPORT FORM NAME: R11
ENTER OPTIONS, M = LEFT MARGIN, L = LINES/PAGE, W = PAGE WIDTH
```

```
PAGE HEADING? (Y/N) Y
ENTER PAGE HEADING: CADMIUM CONCENTRATIONS IN MILLIGRAMS PER
CUBIC METER
DOUBLE SPACE REPORT? (Y/N) N
ARE TOTALS REQUIRED? (Y/N) N
COL        WIDTH,CONTENTS
001        20,PERSONSAMP
ENTER HEADING: NAME
002        10,SSNUM
ENTER HEADING: SS NUM
003        15,JOBTITLE
ENTER HEADING: JOB TITLE
004        10,CONCENTRAT
ENTER HEADING: CONC.
005
```

The REPORT FORM command performs a search of data base and gives the output in the predefined stored format that was generated by the REPORT command:

. REPORT FORM R11 FOR CONTAMCODE = '002'

PAGE NO. 00001
02/14/84

CADMIUM CONCENTRATIONS IN MILLIGRAMS PER CUBIC METER

NAME	SS NUM	JOB TITLE	CONC.
ADAMS,THOMAS	345678901	CRANE OP	0.325
ADAMS,THOMAS	345678901	CRANE OP	0.114
ADAMS,THOMAS	345678901	CRANE OP	0.209
ADAMS,THOMAS	345678901	CRANE OP	0.192
ADAMS,THOMAS	345678901	CRANE OP	0.435
ADAMS,THOMAS	345678901	CRANE OP	0.001
ADAMS,THOMAS	345678901	CRANE OP	0.505
ADAMS,THOMAS	345678901	CRANE OP	0.007
BROWN,FRANK	567890123	CRANE ASST	0.171
BROWN,FRANK	567890123	CRANE ASST	0.038
BROWN,FRANK	567890123	CRANE ASST	0.372
BROWN,FRANK	567890123	CRANE ASST	0.025
BROWN,FRANK	567890123	CRANE ASST	0.038

DBASEII has its own programming language which allows for performing complex data base tasks. Programs are stored as command files. The MODIFY COMMAND instruction permits entry of a program. An example of a simple program follows:

```
.MODIFY COMMAND C3
NEW FILE
SET EJECT OFF
USE IHDATA
ACCEPT 'ENTER PAGE HEADING' TO PH
ACCEPT 'ENTER CONTAMINANT CODE' TO CC
? PH
REPORT FORM R6 FOR CONTAMCODE = CC
```

The program is recorded by depressing the control and W keys simultaneously. This example assumes that a report form R6 had been created earlier.

To execute a program the command DO followed by the program name is issued:

```
. DO C3
ENTER PAGE HEADING:CADMIUM CONCENTRATIONS IN MILLIGRAMS PER CUBIC METER
ENTER CONTAMINANT CODE:002
```

The output is:

CADMIUM CONCENTRATIONS IN MILLIGRAMS PER CUBIC METER

PAGE NO. 00001
02/14/84

NAME	SS NUM	JOB TITLE	CONC.
JONES,JOHN	123456789	CRANE OP	0.018
JONES,JOHN	123456789	CRANE OP	0.016
JONES,JOHN	123456789	CRANE OP	0.020
SMITH,ROBERT	234567890	CRANE OP	0.099
SMITH,ROBERT	234567890	CRANE OP	0.011
SMITH,ROBERT	234567890	CRANE OP	0.071
ADAMS,THOMAS	345678901	CRANE OP	0.325
ADAMS,THOMAS	345678901	CRANE OP	0.114
ADAMS,THOMAS	345678901	CRANE OP	0.209
ADAMS,THOMAS	345678901	CRANE OP	0.192
ADAMS,THOMAS	345678901	CRANE OP	0.435
ADAMS,THOMAS	345678901	CRANE OP	0.001
ADAMS,THOMAS	345678901	CRANE OP	0.505
ADAMS,THOMAS	345678901	CRANE OP	0.007
DOE,PAUL	456789010	CRANE ASST	0.014
DOE,PAUL	456789010	CRANE ASST	0.018
DOE,PAUL	456789010	CRANE ASST	0.032
DOE,PAUL	456789010	CRANE ASST	0.052
BROWN,FRANK	567890123	CRANE ASST	0.171

Information from a data base can be transferred to another using the UPDATE command. Two data bases can be combined to form a third by using the JOIN command:

```
. USE IHDATA
. SELECT SECONDARY
. USE CODE
. JOIN TO NEWFILE2 FOR P.CONTAMCODE = S.CONTAMCODE FIELD
PERSONSAMP,CONTAMCODE,DECODE,CONCENTRAT,JOBTITLE

. USE NEWFILE2
. BROWSE
```

RECORD # :00001

PERSONSAMP	CON	DECODE	CONCENTR	JOBTITLE
JONES,JOHN	001	LEAD	0.052	CRANE OP
JONES,JOHN	002	CADMIUM	0.018	CRANE OP
JONES,JOHN	003	BERYLLIUM	4.595	CRANE OP
JONES,JOHN	001	LEAD	0.050	CRANE OP
JONES,JOHN	002	CADMIUM	0.016	CRANE OP
JONES,JOHN	003	BERYLLIUM	1.983	CRANE OP
JONES,JOHN	001	LEAD	0.033	CRANE OP
JONES,JOHN	002	CADMIUM	0.020	CRANE OP
JONES,JOHN	003	BERYLLIUM	2.069	CRANE OP
JONES,JOHN	001	LEAD	0.026	CRANE OP
SMITH,ROBERT	002	CADMIUM	0.099	CRANE OP
SMITH,ROBERT	003	BERYLLIUM	13.245	CRANE OP
SMITH,ROBERT	001	LEAD	0.029	CRANE OP
SMITH,ROBERT	002	CADMIUM	0.011	CRANE OP
SMITH,ROBERT	003	BERYLLIUM	3.044	CRANE OP
SMITH,ROBERT	001	LEAD	0.141	CRANE OP
SMITH,ROBERT	002	CADMIUM	0.071	CRANE OP
SMITH,ROBERT	003	BERYLLIUM	0.933	CRANE OP
SMITH,ROBERT	001	LEAD	0.075	CRANE OP

In the above example, a file which contained the name of the contaminant which corresponds to each contaminant code was used to create a new file which contained both name and code.

CHAPTER 2

INFORMATION SYSTEM BUILDING BLOCKS

CHARLES W. McNICHOLS, PhD
Professor of Management, Department of Management, Clemson University, Clemson, South Carolina

INTRODUCTION

A systems view of microcomputer support for safety and health professionals is presented as a framework for a knowledge base used to select system elements or building blocks in the hardware, software, people, and procedures categories. The general objectives of computer applications at the Electronic Data Processing (EDP), Management Information System (MIS), and Decision Support System (DSS) levels are introduced and provide a structure for identifying the class of safety and health automation problems to be approached. Potential applications for knowledge-based (expert) systems in the occupational safety and health field are suggested. The importance to information system success of balancing the investment in each of the four building block categories is emphasized. A brief description of specific components in each category is provided, along with parameters useful for their evaluation and selection.

The Four Categories of System Building Blocks

Selecting the proper components for a computer-based occupational safety and health information system requires two major classes of activity: defining the problem, and choosing appropriate tools to solve the problem as it has been defined. A model which classifies these tools is provided by the four-legged support structure for safety and health applications presented in Figure 1. The four legs represent four categories of system building blocks which will be examined in this chapter.

Hardware

Computer hardware consists of the physical elements of the system which may be seen and touched. Two major classes of hardware will be considered: the computer itself, and peripheral devices attached to the computer, generally for input/output purposes.

Software

Computer programs, the lists of step-by-step instructions that are loaded into the computer to allow it to perform each job function, represent the software base.

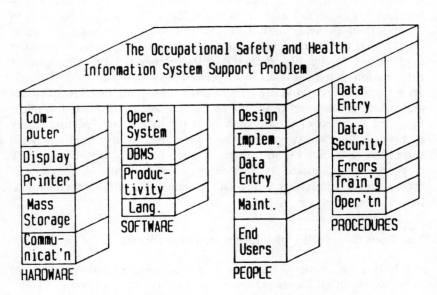

Figure 1. System building blocks: the four categories of support.

People

A combination of special skills and tedious labor is required to implement and operate an information system successfully. A variety of talent is required from the initial design activity through the operational phase of a computer system's life cycle.

Procedures

Documentation, training, provisions for software correction and extension, and providing for system security fall into the procedures area.

The purpose of the four-legged analogy is to emphasize that almost any level of computer support, from none to a multi-million dollar mainframe, can be (and probably has been) applied either successfully or unsuccessfully to almost any safety and health computer support problem. Effective support results from a proper balance among the four support mechanisms: the length of the legs is not as important as making them all about the same length. Consider the following examples:

- Five thousand dollars is budgeted to buy a microcomputer system to be used primarily for letter and report preparation in a plant safety office. To extend use of the machine to other tasks, expensive electronic spreadsheet, business graphics and data base management software is purchased. Because of the heavy investment in software, only an inexpensive dot matrix printer can be afforded. It does not produce correspondence quality printing, so important letters and reports must still be typed. The hardware/software balance is inappropriate for the primary problem the system was to solve.
- A complex corporate-wide integrated system supporting health and safety management data collection and reporting is implemented on a mainframe computer. The system is professionally designed and implemented, well suited to the organization's needs, and is technically perfect. However, little effort is expended on preparing documentation appropriate for inexperienced computer users, and no formal training program is provided. Employees have difficulty using the system, and it is a failure. In this case, the procedures leg of the information systems foundation is inadequate for the complexity of the software and hardware system.
- A minicomputer-based information retrieval system is designed with special features, such as extensive help screens and prompts, to make it user friendly. The computer resources to support the extensive data transmission required by such a system are inadequate when many people try to use it simultaneously. Response time is so slow that

users' needs cannot be satisfied. In this example, the relationship between software complexity and hardware capability is out of balance.

The Systems Concept

A system is a collection of entities which interact to accomplish a goal. For our purposes, the goal is computerized support for safety and health activities. It is interaction among the four categories of system building blocks that makes the "systems" terminology appropriate. Any system can be visualized as a subsystem of a larger system. The output from one subsystem is the input to another. Some of the major input and output relationships among the four system building blocks are portrayed in Figure 2. These interactions motivate the need to examine and balance the four components in choosing a system.

System design determines the software required for implementation, which in turn constrains the hardware to a configuration which will adequately support system activity. Hardware and software decisions

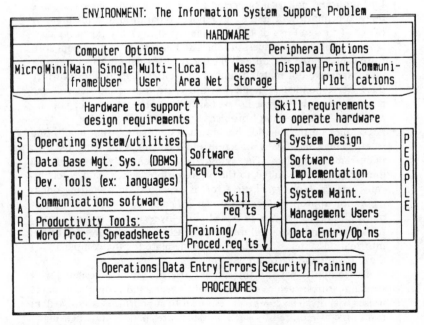

Figure 2. Major interactions among the four support categories.

influence the level of personnel skill required to operate and use the information system. These requirements generate needs for training and operating procedures development. Procedural requirements may also influence required personnel skill levels. For example, the specific methods available in software to provide data security will influence the degree to which the system can be operated without the involvement of full-time information systems specialists. More detailed interactions could be depicted in Figure 2, however, its general point is the need to consider the overall system as well as the individual pieces.

DEFINING THE PROBLEM

Levels of Information Processing Support

In an ideal world, the selection of computer support for safety and health activities would begin with defining in general, and then in detail, what the computer system is to do.[1] With this background, the proper configuration of software, hardware, personnel, and procedural requirements to accomplish support tasks could be specified. In practice this type of design approach is difficult to implement. Many different activities performed in the safety and health field could benefit from computer support. Once experience with an initial system is obtained, its users are likely to think of new applications, some of which may be more valuable than those which led to the initial system design. High demand for additional computer-based support capabilities is often an indication that the system is highly successful, so systems should be designed with extension in mind. In data processing terminology, this is the concept of "evolvability."

Tables I and II summarize responses from a 1985 survey of safety and health professionals concerning their use of computers.[2] Typical applications and a frequency distribution of the number of applications are presented in the tables. Note that few users implement only a single application.

Many of the most valuable and appropriate applications for microcomputer support of safety and health professionals will be decision-oriented rather than clerically-oriented. Information and decision requirements which are not predictable at the time a system is designed, and which support activity where a substantial amount of human expertise is involved, demand flexible, accessible hardware and software tools. Such tools are readily available for microcomputers.

The levels of computer support for information processing are often

TABLE I. Safety and Health Applications of Computers

Sample Size: 49

Application	Percentage* Reporting
MSDS data base	49
Air sampling records	53
Noise measurement	51
Medical records	27
Safety/accident records	35
Engineering control design (ex: ventilation)	14
Other(s)	27

*Percentages add to more than 100% because of multiple applications.

TABLE II. Number of Computer Applications Implemented

Sample Size: 49

Number of Applications	Percentage* Reporting
0	8
1	10
2	24
3	20
4	16
5	14
6	2
7	4

*Percentages add to less than 100% because of rounding.

defined through an EDP, MIS, DSS hierarchy, which we will augment to include the related support areas of office automation and knowledge-based systems (Figure 3). In examining these support levels, note that an integrating factor for applications at all levels is the data resource. Adequate data base design and implementation using a data base management system that permits easy extension are the keys to information systems which can grow to meet new requirements.

ELECTRONIC DATA PROCESSING (EDP)

The automation of manual clerical record keeping represents the oldest business application of computer equipment. A typical EDP project is justified in terms of labor cost savings and improved processing timeliness. Such projects automate very structured, repetitive tasks with com-

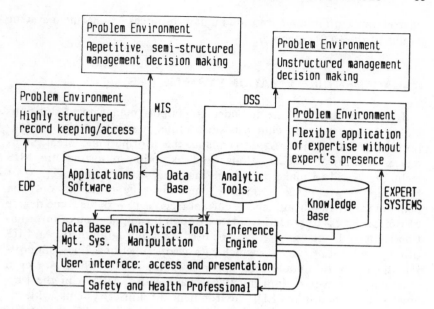

Figure 3. Types of information system support.

pletely predefined inputs and outputs, and most often support a single functional area such as accounting or personnel. EDP applications usually involve the storage of very large collections of data, with a correspondingly high volume of updating transactions. There will be a need for equipment which can input, retrieve, and output data at high speeds with high reliability. Multiuser computer systems, permitting access to a single data base for update and retrieval purposes by many users working at different locations simultaneously, would usually be required. Software will often be implemented through a sophisticated data base management system (DBMS).[3-6] The DBMS consists of a standardized set of tools used by the applications software to manipulate the data base. This speeds application development, standardizes data backup and recovery, and data security procedures. Successful EDP systems require detailed written procedures and supervised operator activity to assure a disciplined approach to data entry and protection. EDP applications are often the first ones implemented when an organization automates, because of the need to come to grips with overwhelming record keeping tasks, and because it is usually easy to cost justify these applications. The payroll application in a corporate human resources management system would fit the EDP category.

MANAGEMENT INFORMATION SYSTEMS (MIS)

Systems at this level are intended to support management decision-making activities, with emphasis on routine, repetitive decision situations. Raw data are processed to enhance the information content. Summaries, statistical extrapolations, and exception reporting are MIS techniques used to make data more meaningful to the manager. Most MIS reports are predefined, but some might be produced only on demand rather than routinely. A current trend is to provide some degree of flexible access to predefined data through the application of information retrieval languages that may be used by nonprogrammers. An MIS often permits access to data which cuts across functional lines, supporting the needs of middle and upper level managers. There are even stronger incentives to implement software with a DBMS in the MIS environment than in the EDP environment. Multiuser systems, with the possibility of terminals directly used by managers, provide the typical MIS implementation vehicle. Data accessed in an MIS will often include the same data used for EDP purposes. The human resource management system used as an EDP example becomes an MIS once it is enhanced to include routine and on-demand reports to aid in managing such activities as employee training, health examination requirements, and applicant-job matching.

DECISION SUPPORT SYSTEMS (DSS)[7,8]

The unstructured decisions and related information requests with which middle managers are often involved are not well supported by inflexible EDP and MIS systems. Solutions to this class of problems are obtained through phone calls, manual searches through filing cabinets, rough estimates, data organized on ruled paper forms, and desk calculator manipulation. A decision support system generator provides the analyst or decision maker with an integrated collection of computer support tools to make work in this unstructured environment more productive. The classical decision support system includes a data base management system which allows the organization, manipulation and retrieval of data, a model base management system which could include simulation and statistical analysis tools, and a user friendly software interface which allows an analyst or decision maker to operate the system without programmer support. Facilities which produce reports and graphs easily are often a part of this interface. The simplest decision support tools are provided by the integrated electronic spreadsheet packages which have facilities for data base management, computations, and report and graph generation. An important element of many of the current microcomputer-based decision support tools is its ability to access data stored in mainframe data bases. Thus, the DSS becomes an integrated part of the MIS and EDP structure. A safety and health relevant DSS might be designed to facilitate epidemiological studies through statistical analyses of a data base containing employee health records and exposure histories.

KNOWLEDGE (EXPERT) SYSTEMS[9,10]

While knowledge or expert systems are most often implemented as self-contained applications, they often are used to approach problems which fit the DSS framework. Knowledge systems go beyond the storage of raw facts by encoding human expertise in a knowledge base and including processing logic that allow them to reason with this expertise, much as a human expert reasons. In the classic DSS definition, we have portrayed the decision maker (the expert) directly interacting with the computer system. The DSS provides data and models to manipulate these data, while the decision maker provides the reasoning and expertise. In the knowledge-based system, some or all of the expertise and reasoning facility is provided by the computer implementation. The major elements of an expert system include a knowledge base, which stores the expertise, and an inference engine, which provides the reasoning mechanism. These systems simulate a consultation with a human expert. A series of questions is posed on the screen, and the user replies to these from the keyboard. The course of the questioning is controlled by the inference engine, software which seeks input to allow efficient resolution of the goals of the consultation: providing advice to the user comparable in quality to the advice a human expert would provide. The most common way to represent expertise in a knowledge base is in the form of rules with the following general appearance:

```
     IF—the incident involves chlorine,
   AND—a spill is likely,
 THEN—evacuate unnecessary people,
   AND—wear positive pressure breathing apparatus,
   AND—wear full protective clothing.
```

A realistic knowledge base could contain hundreds or thousands of rules. The conclusions from one rule (the THEN part) may affect the premises (the IF part) of other rules. The inference engine controls the logical path followed in evaluating the rules to determine what to do next. For example, before the rule shown above could be evaluated, other rules might provide information leading to the conclusion that the toxic substance is chlorine and that the conditions of the incident make a spill likely. The process of encoding expertise so that it may be stored in the knowledge base is called knowledge engineering.

Expert systems have been constructed to do medical and maintenance diagnosis, and to provide advice on topics as diverse as configuring computer hardware and formulating a case for asbestos litigation. Potential safety and health relevant applications include systems to suggest appropriate protective equipment based on exposure circumstances or recommending actions to take when there is a toxic spill. A knowledge-based system can make the expertise of a small number of living experts available inexpensively and simultaneously in dispersed physical locations.

OFFICE AUTOMATION

This subject will not be treated as a separate level of information system support since office automation tasks can be defined at the EDP, MIS and DSS levels. Word processing is the most common office automation project, and can be viewed as a small scale EDP application, where the data are in a text form. Mailing list processing, electronic mail, and meeting scheduling are other applications which have varying degrees of EDP and MIS flavor. In many business organizations, the equipment supporting office automation will be integrated, through communications facilities, with other corporate information processing equipment.

When initiating the automation of safety and health activities it is important to keep the evolvability concept in mind. Most initial automation efforts will be for record keeping. However, an EDP system implemented for clerical support might provide MIS support to middle management if flexible data retrieval and report generation capabilities can be obtained. Statistical and graphical presentation tools which access the same data might provide the flexibility needed to provide meaningful decision support. While all of these possibilities cannot be implemented at once, good initial decisions in the hardware, software, people, and procedures categories will make their eventual realization more likely. It is important to think about automated support for the activities you know you should be doing, not just the activities you do now because of the press of daily tasks. Consider whether the computer capabilities which will do most to make you effective in your profession are at the EDP, MIS or DSS level. Finally, when implementing record keeping activities on a microcomputer, consider the likely expansion of these activities and the extent to which, as a safety and health professional, you can justify spending your time running an EDP operation, with the attendant problems of operator supervision, hardware maintenance, data preparation, and error recovery.

The Fundamental Tradeoffs (There Is No Free Lunch)

There are some unfortunate (and often unstated) hardware and software tradeoffs which may lead to frustration in microcomputer system implementation. Figure 4 indicates system attributes which have natural tendencies to pull in opposite directions. Cost considerations should not be overemphasized, since the cost of microcomputer hardware and software is low relative to the people costs that will be involved in its use. More important is the idea that the system which is easiest to get started with may ultimately prove inadequate.

The difference between a list of *each* of the things a microcomputer

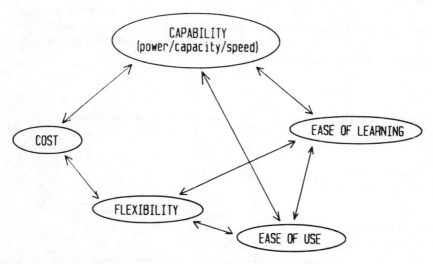

Figure 4. Fundamental tradeoffs in microcomputer systems.

system can do, and a list of *all* of the things it can do simultaneously can be an important one. Microcomputer-based information systems can be powerful or friendly or flexible or fast or provide multiuser capabilities or provide inexpensive dedicated productivity support or implement sophisticated data security provisions or run off-the-shelf software. However, a microcomputer system will not deliver all of these capabilities at one time.

A Knowledge Base For Evaluating Alternatives

In the remaining sections the four categories of system building blocks will be examined in more detail. This material consists of lists and brief descriptions of options, parameters defining capabilities, and advice which ties attributes of the problem to suggested capabilities, and thus to recommendations for components to acquire. This material has been approached as a knowledge engineer might initially approach the task of automating a consultant's specific system implementation advice, augmenting the lists of major alternatives with general rules for making appropriate choices. Selections in each of the four areas must be made in the context of the level of decision support required.

All of the discussion beyond selection of the computer itself assumes implementation of a microcomputer-based system. Because of its overwhelming popularity in the business world, the IBM-PC and compatible computers form an implicit basis for consideration of hardware and software alternatives. Table III presents data from the survey mentioned earlier,[2] which indicates the pervasiveness of the IBM-PC for occupational safety and health support.

TABLE III. Microcomputers Used by Safety and Health Professionals
Sample Size: 49

Computer Used	Percentages* Reporting
IBM-PC or XT or AT	73
IBM-PC compatible	10
Apple II or III	10
Apple Macintosh	6
Radio Shack	8
Other	31

*Percentages add to more than 100% because of multiple computers.

HARDWARE

The Computer Itself

Level Of Computer Support

Microcomputer: Technically, any machine which uses a microprocessor integrated circuit for its central processor is a microcomputer. It is generally small, desk-top or portable, and most are appropriate for the support of a single user.

Minicomputer: Most minicomputers are almost always multiuser systems (allowing simultaneous access by users working at terminals) with more powerful input and output processing capabilities than microcomputers.

Mainframe: A large scale computer system which can support many users at one time is a mainframe. It can rapidly access very large data bases and work with the widest variety of peripheral devices.

Microcomputer Considerations

Compatibility: The IBM-PC standard is significant because of peripheral and software availability, and corporate standardization. Two kinds of compatibility are relevant—the ability to use hardware (usually circuit boards) designed for the IBM, and the ability to run software written for the IBM.

Number of Simultaneous Data Base Users: A multiuser access requires at least a PC-AT level machine or local area network.

Random Access Memory (RAM) Capacity: The number of characters of data or software the computer can work with internally. General advice—buy full memory capacity at the outset (it is inexpensive and useful for both speed and convenience).

Operating Speed: For applications with extensive calculations (statistical analysis, big spreadsheets) IBM-PC compatible machines based on the Intel 8086, 80186, or 80286 microprocessors (rather than the 8088 used by the standard IBM-PC) or with higher clock speed may be useful.

Parameters Defining Computer Operating Speed

Width of Data/Computation Path: Individual positions in the binary representation of data or program instructions are measured in bits. Microcomputers are usually 8-, 16-, or 32-bit machines. The number of bits manipulated at once in moving data and performing computations influences speed. Sixteen-bit microcomputers are most common. True 16-bit microprocessors include the 8086, 80186, and 80286, but not the 8088 used in the IBM-PC, which performs 8-bit data transfers.

Clock Speed: The same microprocessor chip may be operated at varying speeds. The standard clock speed for the IBM-PC and XT is 4.77 mHz. Higher clock speeds lead to faster operation.

Math Coprocessor Option: The 8087 math coprocessor (80287 for the PC-AT) speeds up numerical calculations (but not input/output or other data manipulation operations). Software specifically designed to make use of this hardware feature is required to gain a speed advantage.

The Issue of Input/Output vs. Internal Speed: The operating speed of many computer applications depends more on the speed of data transfer (to and from disk storage, and to a printer) than on numeric calculation speed. An accurate comparison of different computers' throughput can only be accomplished by performing a typical mix of operations using real or simulated data (a process called "benchmarking").

Microcomputer Peripheral Devices

Display Considerations

Text Characters and Graphics: These are produced from tiny dots called "pixels" which are selectively illuminated to form the pattern representing a character or drawing. The "resolution" of a display is defined by the number of pixels displayed across each row (horizontal resolution), and the number of rows (vertical resolution).

Resolution (Horizontal/Vertical Pixels): Normal IBM options (for the Color Graphics Adapter or "CGA") provide a choice of 320 horizontal by 200 vertical or 640 horizontal by 200 vertical pixels. Four colors can be displayed in the 320 by 200 mode, and two in the 640 by 200 resolution. The color count includes the background color.

The Display Subsystem: The combination of a display adapter (which is a circuit board installed inside the computer case) and a monitor forms the display subsystem. The circuit board and monitor must be matched electronically (Figure 5 and Table IV).

Monochrome Monitors: These display a single color of text or graphics on a background color which is usually black. Color monitors usually produce their displays by mixing red, green, and blue colors, and an intensity signal.

Digital vs. Analog Display: The least expensive monochrome or color monitors are driven by a composite (analog) signal. These systems are identifiable by the use of RCA plugs (audio jacks) on the cable connecting the monitor to the graphics adapter. Monochrome quality is ade-

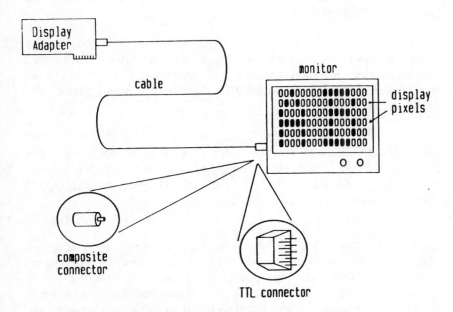

Figure 5. Display subsystem components.

TABLE IV. Popular IBM-PC (and compatible) Display Configurations

Adapter	Monitor	Text Cols/Rows/Quality	Graphics Hor/Vert Pixels/Colors
IBM Mono	Mono TTL	80/25/Good	None
Hercules	Mono TTL	80/25/Good	720 × 348/2 (monochrome)
IBM CGA	Mono composite	80/25/Poor	320 × 200/2, 640 × 200/2
	Color composite	40/25/Awful	320 × 200/4, 640 × 200/2
	Color RGB	80/25/Poor	320 × 200/4, 640 × 200/2
IBM EGA	Mono TTL	80/25/Good	640 × 350/2 (monochrome)
	Color RGB	80/25/Poor	320 × 200/16, 640 × 200/16
	Enhanced RGB	80/25/Good	320 × 200/16, 640 × 350/16

quate, while the color quality is generally inadequate with these systems. Twenty-five lines of 80 characters will be readable on a composite monochrome monitor, but only 40 characters per line can be viewed comfortably on a color composite monitor. Digital displays (the IBM monochrome monitor and "RGB" [Red/Green/Blue] color monitors) are most widely used in IBM-PC systems. Digital monitors are often referred to as "TTL" (transistor-transistor logic) devices. A nine-pin connector on the cable identifies digital systems.

High Resolution Monochrome Graphics: The Hercules graphics card and compatible products produce high resolution (720 × 348 pixel) graphics images on monochrome TTL monitors like the standard IBM monochrome monitor.

High Resolution Color Graphics: Provides up to 16 colors with 640 horizontal by 350 or 400 vertical pixels. An example of a high resolution color graphics subsystem is the IBM Enhanced Graphics Adapter (EGA) used with the enhanced color display.

High Quality Text: This is needed for word processing. The IBM monochrome monitor produces characters in a 9 (horizontal) by 14 (vertical) pixel matrix. Text characters produced by the CGA in either monochrome or color are in an 8 × 8 pixel matrix, and are of noticeably lower quality. High resolution color adapters like the EGA produce good quality characters using an 8 × 14 or 8 × 16 matrix.

Software Support: To have access to a large collection of graphics software, you need to buy hardware which is compatible with that for which most software is designed. Most available graphics software supports the IBM-CGA. Many programs support the Hercules graphics format on the IBM monochrome monitor. The emerging standard for high resolution color graphics is provided by the IBM Enhanced Graphics Adapter.

Cost: High resolution color graphics more than doubles the cost of the interface board and monitor compared to the cost of a standard CGA and RGB monitor.

Data Entry Considerations

Keyboard: Alternate keyboards may be purchased which offer features such as cursor keys separate from the numeric keypad and lights to indicate whether the keyboard is in upper or lower case shift.

Light Pen: Allows identification of a screen position by pointing at an illuminated screen area with a pen-like object. Light pens are effective for operator actions that involve pointing, but not very effective for drawing.

Mouse: A pointing device used as an alternative to cursor (arrow) keys for moving the cursor (character indicating current screen position) on the screen. A good interface for drawing or moving objects on the screen.

Touchscreen: Allows for the selection of options or identification of screen position for other purposes by pointing at the screen location.

Touchpads: Similar to touch screen, but a tablet separate from the video screen is employed. May be used for option selection or digitizing positions from a chart placed on top of the tablet.

Voice: Voice recognition devices permit option selection by recognizing words or short phrases from a fixed vocabulary when the device has been trained for a specific operator.

Bar Code Readers: Useful in data entry situations where keyboarding needs to be avoided because use of a keyboard is impractical or would lead to inaccurate input. Bar code readers typically simulate computer keyboard input.

Digitizers: Used to capture the coordinates of points on a drawing for computer input. Some plotters can be used as digitizers.

Analog to Digital Converters: Used to convert continuously varying electrical voltages into a numeric (digital) form which can be processed by the microcomputer. It is useful for direct input from laboratory or other measuring equipment-like spirometers.

Portable Computers (Serial Interface): These are small portable computers which can be used as data gathering devices, then transmit their data to a desk-top machine through a serial interface. Bar code readers and analog to digital devices are often attached to these computers for remote data gathering.

Parameters Governing Use Of Alternate Input Devices

Type and Source of Data: Considerations such as the form of the data (text, numeric, graphic description, analog) and the location of the data (lab, shop floor, remote sensor) are relevant to selecting alternate input devices.

Skill, Training, Location of Data Entry Personnel: Typing and other clerical skills as well as the physical circumstances under which data are gathered must be considered in choosing an appropriate data entry mechanism.

Software Support: Custom software is often required to deal with specialized interfaces. Programs that will drive the device are sometimes provided with the hardware. Many devices simulate keyboard data entry or provide data through a serial (communication) port. This may simplify custom software development, since BASIC's serial input and output capabilities can be used.

Cost: Software development, special training, and the possible need for custom electronics interfaces should be considered.

Mass Storage Considerations

Floppy Disks: The least expensive mass storage mechanism is the floppy disk. Advantages are low cost and ease of software and data transfer. Two floppy disk drives are recommended for any microcomputer installation to simplify floppy disk copying. The IBM format stores approximately 360,000 characters (bytes) of data on each disk.

Hard Disk: The most common rigid or hard disk drives store 10 or 20 megabytes (million characters or bytes) of data. Removable (cartridge) hard disks are available at prices currently about five times the price of a fixed hard disk. A backup device is needed to secure the investment in data entry. Streaming tape (digital recording on a tape cassette cartridge) is the most common approach, but a second hard disk or interface to a video cassette recorder are alternatives.

The Future: An emerging technology with potential relevance to safety and health applications is mass storage using video disk or Compact Disk Read Only Memory (CDROM). Very large volumes (hundreds of millions of characters) of data and digitized pictures may be stored in a read-only format, then randomly accessed under computer control. Inexpensive distribution of high volume data such as material safety data sheets might be implemented with these media.

Printer Considerations

Technical Approaches to Printing: Printers used with microcomputers produce one character at a time by moving a printhead across the paper. *Dot matrix* printers form characters by firing the proper combination of a vertical column of pins at each column position as the printhead increments across the page. A *Daisy wheel* or thimble printers produce fully formed characters much like a typewriter. A wheel containing all of the characters on spokes around its circumference turns until the desired character is over the print position, then a hammer hits that type-bar against a ribbon. All daisy wheel, and most dot matrix, printers are impact printers. They produce characters by causing the print mechanism to strike a ribbon. Alternatives include *thermal* and *electrostatic* printers, which produce characters on special paper using a dot matrix mechanism with electrically charged pins. *Ink jet* printers spray ink droplets on the paper, again through a mechanism that is of the dot matrix variety. For high speed, high quality printing, laser printers use a variation of a photocopying mechanism to produce text and graphics with a very high resolution (many dots per square inch).

Quality: For serious word processing applications, the ability of a printer to produce office typewriter ("letter" or "correspondence") quality may be important. Dot matrix printers which are less expensive and faster than daisy wheel printers may have a "near letter quality" mode which produces output quality good enough for most applications. If very high quality output is required, do not overlook the fact that only daisy wheel printers can use a carbon ribbon.

Speed: The speed of character-at-a-time printers is measured in characters per second. Dot matrix printers usually have draft and near letter quality modes, with higher speeds possible in the draft mode. Advertised printer speeds are generally 20 to 50% higher than actual average output speeds because of time lost in advancing the paper.

Paper Handling: Daisy wheel printers normally have friction paper feed mechanisms like typewriters, which are ideal for handling one sheet of paper at a time. A tractor feed mechanism can be added to these printers to allow for continuous forms handling. Sheets or forms are connected by perforations and pulled through the printer using holes punched along the edges. Automatic sheet feeding mechanisms are also available for many daisy wheel printers. Dot matrix printers generally have a pin feed mechanism (pins on the ends of a platen which engage the forms perforations) for use with continuous forms, and may also have friction feed.

Paper Types: Daisy wheel and impact dot matrix printers use standard paper products. Thermal and electrostatic printers use expensive special paper. Ink jet printers generally produce better output with paper that has special ink absorption qualities. Special clear acetate is now available to allow impact and ink jet dot matrix printers to produce overhead transparencies.

Graphics Output Considerations

Printers: Most dot matrix printers will produce black and white graphics output. There are color ink jet and impact dot matrix printers which will produce graphics. There is no standard for graphics printing; one must make certain the software will produce graphics output for the printer purchased.

Plotters: For high quality visual aids, multiple-pen plotters are available. These will produce multicolor graphs and text, changing pens under software control. They can plot on paper or, with special pens, on clear acetate for transparencies.

Cameras: To aid in photographing the screen to produce 35-mm slides, special fixtures which hold the camera and shield the lens from extraneous light are available. Self-contained devices which accept computer graphics program output, and produce high resolution slides are also available. The latter solution requires software capable of operating with the slide producing equipment.

Communications Considerations

Media Exchange: If enough time is available, the least expensive way to move large quantities of data among geographically separated computers is generally to mail floppy disks. This requires compatible floppy disk formats for the computers exchanging data.

Modems: A modem (modulator-demodulator) converts voltages representing digital data into audible tones for transmission over telephone lines, or from these audible tones back into the digit form needed for computer input. By using modems at both ends of a phone connection, two computers can exchange data, or a microcomputer can be used as a terminal to a larger computer. *Asynchronous* modems are the most common type used with microcomputers. They exchange data at speeds of 30 to 240 characters per second, and the timing of data transfer is critical only during the transmission of each character. *Synchronous* modems transfer data in fixed length messages at higher speeds, and at higher costs, than asynchronous modems.

Terminal Emulator Boards: A circuit board installed in an IBM-PC or equivalent, along with special communication software, allows it to emulate the special terminals used with IBM 3270 terminal controller systems or with the IBM system 34/36/38 series of minicomputers. With these emulators, it is possible to switch ("hot key") between running a PC application program and using the computer as a terminal. It is also possible to exchange files between the microcomputer and mini- or mainframe computer.

Protocol Converters: As an alternative to terminal emulator boards, a mainframe computer can be equipped with a protocol converter which permits telephone access from microcomputers equipped with an asynchronous modem and communication software.

Local Area Networks: Local area networks allow the physical interconnection of microcomputers to exchange data and share data bases stored on a central hard disk and to make common use of peripherals such as printers. An interface (called a "gateway") may be provided to allow data transfer between local area networks or between a local area network and a minicomputer or mainframe.

General Hardware Selection Guidelines

If the primary application is at the EDP or MIS level, then data storage, input transaction, and data access/report output volumes will drive the hardware selection. Data bases which include tens of thousands of records with hundreds of daily transactions are good candidates for minicomputer and mainframe implementation. A significant issue is whether data base access must be provided for multiple users. If only a few users need multiple access (say 3 or 4), a high-end multiuser microcomputer provides a possible implementation environment. If many users need access to the same data, a local area network or larger scale computer system is required.

For extensive text and data entry, a display system which produces a high quality monochrome text display is important. If word processing with typewriter quality output is needed, a daisy wheel printer is suggested. Dot matrix printers provide a faster and more flexible output alternative for text and graphics.

If flexible data access and manipulation of the DSS variety will be required, microcomputers with color graphics capabilities are best. Additional output devices such as plotters and cameras will facilitate the use of data in management presentations. If high quality graphics, or a combination of graphics capability and good text quality is required, a high resolution monochrome or color graphics option is suggested.

If a combination of large scale EDP and MIS applications is required along with DSS support, microcomputers capable of downloading data from a larger computer's data base for local manipulation should be considered.

SOFTWARE

The Software Subsystem

Operating System (OS)

Basic Operating System Terminology: The operating system is computer software which allows the user to load and run *application programs*, the software that accomplishes useful work like storing and printing an MSDS. It provides standard routines to operate peripheral devices such as the keyboard, display, and disks. The OS also includes *utilities* that provide functions like listing the files that are stored on a disk or copying data or programs from one disk to another. *Single tasking* operating systems like PC-DOS allow only one program to be loaded for operation at a time. Thus, one person may run one application at a time. *Multiple tasking* operating systems provided by some of the windowing extensions to PC-DOS (Top View, GEM, Windows, Concurrent PC-DOS) permit several applications (e.g., a word processing program and a spreadsheet) to be loaded into memory at the same time. The computer alternates among the tasks so they all appear to be progressing simultaneously. *Multiple user* operating systems such as Unix/Xenix, Pick, or Concurrent PC-DOS allow for multiple tasks supporting multiple users, with users beyond the first working from terminals that have a keyboard and display screen but no local computing capability. Again, the microcomputer alternates among the tasks.

Significance of Operating System Selection: Prewritten software (word processors, spreadsheets, data base managers, etc.) will operate only in conjunction with a specific operating system. One cannot buy a word processing package written for PC-DOS and run it under the Unix operating system, even though both PC-DOS and Unix operating systems are available for the same microcomputer.

Limitations of Multiple Tasking/Multiple User Systems: Multiple tasking/user extensions to PC-DOS are likely to operate poorly with software not specifically designed for them. PC-DOS programs which do extensive text manipulation (like word processors) or produce graphics usually bypass the operating system display routines when they generate output. These programs will not work properly in multiple tasking/multiple user environments. A second problem occurs with file sharing — when multiple applications or users have access to the same data, some way must be found to "lock" an individual record while it is in use for update purposes, because simultaneous updates to the same record will result in data inconsistencies. A data base management system specifically designed for multiple users, working with a multiple user operating system, provides the best solution to the file sharing problem.

Local Area Network Support/Multiuser Support From PC-DOS: The latest versions of PC-DOS provide support (including record locking) for local area networking. There is substantial evidence that multiuser versions of PC-DOS will eventually become available. The advantage of a networked or multiuser version of PC-DOS over Pick or Unix is continued availability of the many PC-DOS based programs which will only run with this operating system.

Communications Support

Data Communications/Terminal Emulation: Communications programs provide for the transfer of data between computers, using the modems described earlier, and telephone lines. They also allow use of a microcomputer as a terminal to another computer. You must match the communication software to the modem and operating system you will use. If you will be using the microcomputer as a terminal to access an information service or bulletin board, you may find a software package that has specific capabilities to simplify access to that service. For example, automatic entry of passwords or user identifiers.

Productivity Enhancement and Application Implementation Vehicles Other Than Programming Languages

Electronic Spreadsheets: The computer analog of a paper form ruled with rows and columns, an electronic spreadsheet allows the entry and manipulation of numeric and text data. Many spreadsheets provide for graphics output and include limited data management capabilities. Some include word processing facilities. These "integrated" spreadsheets are suitable for small scale decision support applications.

Data Base Management Systems (DBMS): A Data Base Management System (DBMS) is a collection of software tools which permit definition of data file formats, and provide facilities for data entry and retrieval. A DBMS allows integration of data across files. More sophisticated DBMSs have a built-in programming language, allow for multiple user file access, and have data security provisions.

DSS Building Tools: These tools are available to allow implementation of DSSs too complex for electronic spreadsheets. Many of these packages have a heavy statistical or operations research flavor.

Expert (Knowledge) System Building Tools (Shells): An expert system building tool provides facilities for knowledge base construction and maintenance, and an inference engine to support consultation with the knowledge base contents.

Programming Languages

General Classes of Languages: Interpretive languages store the program in memory, converting each of the programmer's instructions (called source code) into the detailed machine code ("object code") required by the computer at the time the program is run or "executed." BASIC is typically used in an interpreted mode, and many of the data base management systems such as dBASE III have built-in interpreted languages. *Compiler* languages perform a one-time conversion of the programmer's description into object code. The object code version of the program is all that is required to run the application. Fortran, COBOL, Pascal, and C are usually compiled. The advantage of interpreters is their ease of use—a mistake in the program can be quickly corrected and the program rerun because it stays in memory. However, interpreted programs run more slowly than those which are compiled, so most production applications are compiled. Compilers are available for many interpreted languages, including BASIC and dBASE III, so it is possible to develop software in the interpreted form, then compile it for its final implementation.

General Software Selection Guidelines

If a single user microcomputer is required, PC-DOS will be the operating system of choice because of the large base of prewritten software. If a microcomputer with the maximum amount of random access memory is available, windowing operating environments (e.g., GEM, Windows) can provide a simplified user interface.

The operating system decision is difficult if multiple users must access a common data base. If a data-base oriented system will be developed from scratch, a Unix implementation offers the possibility of easy migration to larger scale computers, together with a long history of successful multiuser application implementation. If data will be accessed infrequently by multiple users who need other PC-DOS applications, a local area network implementation is suggested.

Implementation of any safety and health application which involves information storage and retrieval, for ad hoc inquiries or reports, should be in a data base management system environment. A DBMS which allows multiple user access, or has enhanced versions which would allow multiple user access to be implemented later, is recommended, again to provide for application evolvability.

PEOPLE

System Development/Implementation Skills

The Key—Sound Data Base Design: To build a safety and health information system which will support a variety of current needs, the data base must be carefully designed. Although beyond the scope of this paper, there are many good sources for information on this topic.[3-6] Good designs require continuous involvement of the end users of the system.

Costs of Conversion: An easily overlooked cost of a system results from moving manual records into the automated environment at startup time. This can involve extensive clerical labor which must be provided by employees who already have full-time duties.

Data Entry/Operator Skills

Who, What, When, and Where: An information system is no better than the quality of data it contains. The data preparation, entry, and validation process governs this quality. If full-time data entry clerks will not be employed, who will actually do the data entry? In what form will the data be provided for keyboard entry? Will it be possible to get data entered in a timely enough fashion that the computer data base will be accessed for current information rather than the manual records? How will the data entry process be supervised and how will quality control be maintained? If a single user microcomputer provides the system environment, will the computer be free enough hours every day for data entry activities?

System Maintenance

Evolvability: We have already presented the case for system evolvability, but under the current topic it is appropriate to emphasize the need to provide people support for this evolution. If the system is developed by a consultant or contractor, will this source be available for enhancements? If developed in-house, perhaps by end users, will (and should) these users have the time to extend the applications when needs change? Such issues are relevant even at the level of spreadsheet applications.

End Users

The Video Game Syndrome: While microcomputers can be powerful tools for productivity enhancement, they can also distract managers from managerial activity.[11] The degree of involvement of safety and health managers in application development should be considered to make sure the level is appropriate given the demands of other duties.

General People-Oriented Guidelines

Once microcomputer applications expand beyond small scale personal productivity enhancement such as word processing and electronic spreadsheet manipulation, support from the corporate information system staff or an outside consultant is recommended. End users need to be involved in system design and implementation, but not necessarily to the extent of performing all of the technical activities.

For applications which require construction and maintenance of a large data base, the people costs of conversion and data entry must be considered. In microcomputer systems, these may far exceed the cost of hardware and software.

PROCEDURES

The Need For A Disciplined Approach

Because computer software has a more limited ability to adapt to changing conditions than humans, less flexibility in data input and processing is tolerable in automated systems than in manual ones. The discipline required to operate an automated information system successfully is derived from training, well documented procedures, and supervision.

Documentation Considerations

Requirements:

The level of system sophistication, the number of different users, and the projected lifetime of a system all influence the documentation requirement. A spreadsheet used to calculate exposures from pump readings and lab data will not require as sophisticated a collection of documentation as an integrated accident, sampling, and medical record data base. Documentation is most often produced in a written form, but could include help screens or other coaching devices implemented as an integral part of the application. Some types of documentation to consider include the following.

Software Maintenance Documentation: Details of the design of the application will support error correction and application extension. File designs, formulas used in computations, and a description of program logic would be included.

Data Dictionary: A data dictionary describes each raw fact stored in the data base, providing a narrative definition, source, range of legitimate values, and coding information.

Data Preparation and Entry: Defines how data will be translated from source documents for keyboard entry.

Application Operation: Describes files used by the application, along with expected operator inputs and a list of outputs which should be produced.

Error Recovery: Could be part of the application operation document. What should be done if a disk file cannot be read, the printer runs out of paper, or the power fails in the middle of a two-hour sort?

Training Considerations

User Involvement in Development: When software is developed professionally, involvement of the end users with the design and testing process will provide a cadre for on the job training.

Conversion as Part of the Training Process: Conversion from a manual to automated system typically involves actual end users working with real data. Conversion automatically becomes part of the training process. Availability of documentation and training assistance at this point is critical for successful conversion to a complex system.

New Users: Major software systems are likely to have lifetimes extending beyond the job tenure of people in the safety and health operation when the applications are first implemented. Provisions for teaching new employees to use established applications should be considered. The availability of good documentation is helpful in this function.

Data Security Considerations

Limiting Access to Data: Some elements of a data base may need to be protected because of corporate confidentiality or personal privacy considerations. Mechanisms for protecting data from unauthorized access are relatively unsophisticated in microcomputer systems. They include password-protected access to software which can examine the data base, and physical protection of data by using removable media such as floppy disks or cartridge hard disks. The only meaningful protection for information located on accessible media is provided by encrypting (coding) the data. If data are not encrypted, readily available and easily used utility programs can be employed to directly manipulate the disk files.

Data Backup and Recovery: All mass storage devices are subject to hardware failures, and operator or software problems may lead to destruction of a data base. Backup file copies will avoid the cost and time delays necessary to reenter voluminous data. Procedures must be established to make sure files are routinely backed up, and additional procedures must be available describing how to recover files after a failure.

General Procedures Guidelines

The extent of procedures required is driven by the sophistication of the application, the number of people who will work with the software, and the anticipated system lifetime.

If software (even software implemented with an electronic spreadsheet) will be used by more than three or four people who work closely together, written documentation is suggested.

If a moderate to large data base (more than 1000 records) is involved, written procedures and management activity are required to make sure backups are performed as scheduled.

Software which is developed by outside contractors or which will be used for more than a year or two should be accompanied by software maintenance documentation.

CONCLUSION

Assembling a microcomputer-based system for safety and health support requires consideration of the interaction among system building blocks in the hardware, software, people, and procedures categories. The analysis of immediate needs for computer support should be extended to include consideration of potential future applications, so that the safety and health information system may evolve. It is particularly important to implement information systems which will ultimately support decision-oriented management activity, rather than just automate clerical record keeping. The keys to successful information system evolution are sound data base design and an implementation environment which permits integration of the microcomputer system with other corporate data processing facilities.

REFERENCES

1. McNichols, C.W. and Clark, T.D. 1983. *Microcomputer-Based Information and Decision Support Systems for Small Businesses: A Guide to Design and Implementation*. Reston Publishing Co., Reston, VA.
2. McNichols, C.W. and Schultz, S.A. Survey of individuals who requested a paper presented at the 1985 American Industrial Hygiene Association national meeting.
3. Date, C. J. 1983. *Database: A Primer*. Addison-Wesley, Reading, MA.
4. Martin, J. 1976. *Principles of Data-base Management*. Prentice-Hall, Englewood Cliffs, NJ.
5. Martin, J. 1983. *Managing the Data Base Environment*. Prentice-Hall, Englewood Cliffs, NJ.
6. McNichols, C.W. 1984. *Data Base Management with dBase II*. Reston Publishing Co., Reston, VA.
7. Schultz, S.A., McNichols, C.W., and Kotrla, C.J. 1986. A Prototype Microcomputer-Based Decision Support System for Industrial Hygienists. *Am. Ind. Hyg. Assoc. J.* 47:124–134.
8. Sprague, R.H., Jr. and Carlson, E.D. 1982. *Building Effective Decision Support Systems*. Prentice-Hall, Englewood Cliffs, NJ.
9. Harmon, P. and King, D. 1985. *Expert Systems—Artificial Intelligence in Business*. John Wiley and Sons, New York.
10. Waterman, D.A. 1986. *A Guide to Expert Systems*. Addison-Wesley, Reading, MA.
11. Strehlo, K. Feb. 1986. The Video Game Syndrome. *PC World*, pp. 65–70.

CHAPTER 3

SYSTEMS IMPLEMENTATION

HOWARD L. KUSNETZ, CIH, PE

Manager, Safety and Industrial Hygiene, Shell Oil Company, Houston, Texas

INTRODUCTION

Systems implementation means which hardware and software should be purchased. First is the need to define what the software is to do and work toward the hardware from there. Although the bulk of the discussion will be related to microcomputers, it will be useful to see how Shell Oil entered a mainframe industrial hygiene and health surveillance system program in order to look at the techniques for determining necessary software and hardware.

When Shell set out to build a health surveillance system, it was determined that the system must support seven major activities:

- Epidemiology studies
- Detecting workplace health conditions which require direct dimension
- Medical surveillance programs
- Complying with record keeping and reporting mandates
- Providing data for litigation defense

Note: Opinions, software, and hardware suggestions are strictly those of the author. They must not be construed in any way as being recommendations of, or endorsed by, the Shell Oil Company.

- Public understanding programs
- Input to propose legislative or regulatory activities

From this, nine system requirements were defined:

- Large storage capacity
- Confidentiality
- Operability
- Quality assurance
- Flexibility
- Service levels: availability, reliability, maintainability
- Cost effectiveness
- Minimum location impact
- Retrievability and display

Once the requirements were identified, the initial system components could then be specified. There were six components or modules in the system:

- Biometrics
- Morbidity data
- Mortality data
- Personal attributes
- Work history
- Exposure history

Of these six, the first three were very definitely in the medical area. The last two were in the industrial hygiene area. The personal attributes bridged the two.

It was evident, therefore, that a massive computer system was needed. However, it was also evident that such a system was not inexpensive. The company's costs were in the million dollar range. Most agencies need much less. This is typical of small units, especially governmental industrial hygiene offices. This chapter will address the question of low price and availability and make the following assumptions:

1. Industrial hygiene units have rather limited budgets.
2. Individual users are being discussed, and networking at this point is not indicated.
3. It is recognized that the IBM-PC, the PC/XT, the PC/AT, and the microsoft PC-DOS and MS-DOS are de facto standards. This means the software and hardware must be IBM compatible, but does not mean they must be IBM equipment. However, given the tremendous base of programs written for the IBM, any system that will not read off-the-shelf software is not likely to be useful.

Five basic uses of software to be considered are (1) it must handle data, (2) it must display data relations, (3) it must be able to be used for writing reports, (4) it needs to be able to support making presentations, and (5) it must be able to communicate, particularly over long distances.

Data handling refers to spreadsheets and data bases, i.e., huge collections of records. Data bases may be flat file or relational. Flat file refers to a single set of data; relational refers to mixed sets of data, each set of which has a component in common with every other set. For example, an industrial hygiene sample file data would include an individual's name and social security or employment number. The work establishment employee record file would also include the name and the social security or work number. These common elements in a relational data base system mean that much other data about the individual, his date of birth, address, or work history, do not have to be repeated since they can be accessed from one file and utilized in another.

Data relations means the ability to represent graphically the relations between data.

Report writing comprises text editing and word processing. Text editing is a simplified method of word processing in which short notes are written. Word processing is a more detailed concept which includes special formats, printing characters, use of columns, and so on.

Presentations refer to printed graphics, overhead transparencies, or 35-mm slides.

The minimum hardware comprises a hard disk with at least a 20-megabyte hard disk and 640,000 bytes (640K) of RAM. There should be plenty of empty slots on the mother board for inserting special boards such as expanded memory, color drivers, or communications. The system should have at least one, and preferably two, serial ports, a parallel port, and graphics capability.

The IBM family of computers will certainly support all these requirements. But to a much greater extent, the computer press and advertising are full of equipment that come very close to IBM capability such as the Leading Edge Model B, the Epson Equity, and the PC's Limited Turbo PC. The latter, for example, was advertised in the March 25, 1986, *PC Magazine*. It is a PC compatible with a 640K RAM, 135-watt power supply, 2-speed clock, 8 slots, 2 half-height floppy drives, 20-megabyte hard drive, multiple function card with serial and parallel ports, clock and calendar, monochrome graphics card, and monochrome display for a total of $1780. Similarly priced bargains are available just by looking through the popular computer press.

Software supplies can be purchased at retail, but huge savings can be obtained through mail order purchase. If commercial programs are not wanted, then shareware and freeware are also available through PC user groups (PCUGs) and special interest users groups (SIGs).

Freeware are programs that have been put in the public domain. Many are good, but many are untested and untried. It is up to the individual to see which fit. Shareware are copyright programs which the author will permit one to copy providing the program is unaltered and full credit is given. In the shareware program the author asks that if the program is useful, some sum, usually less than $50, be sent. Under those circumstances, a very adequate library of programs can be obtained for very little money.

There are many specific programs. First, consider spreadsheets. A spreadsheet is the most versatile program available for industrial hygiene use. Of the spreadsheets, Lotus 1-2-3 is a de facto standard. This is because its user base is immense. Help groups are available in almost every major city. There are many templates or applications programs available for the asking through users groups and from Lotus itself. There are also many other sophisticated template programs that are available commercially. The next few illustrations show Lotus programs and the manipulation of data. Illustrated are a series of tables showing samples of dimethyl chicken wire (DMC). The compound is shown in Figure 1. Sample results, as they are first listed, are shown in Figure 2.

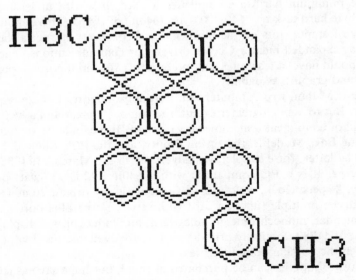

Figure 1.   Molecular structure of di-methyl chicken wire.

```
SAMPLE DATA - RESULTS ONLY
 #    Date       Employee          ID           Mtrl    Conc   Unit  TWA  PPE
 1  02/12/84  Jones,  D F      253-84-9374  DMC      12.35  ppmv   Y    N
 2  02/16/84  Green,  J L      976-52-3476  DMC      23.91  ppmv   Y    N
 3  03/09/84  Jones,  D F      253-84-9374  DMC      19.63  ppmv   Y    N
 4  03/09/84  Smith,  T R      398-45-2741  DMC      19.63  ppmv   Y    N
 5  03/22/84  Smith,  T R      398-45-2741  DMC       8.50  ppmv   Y    N
 6  06/19/84  Jones,  M V      754-28-3519  DMC       2.81  ppmv   Y    N
 7  08/31/84  McGillicudy A O  358-82-1473  DMC      33.54  ppmv   Y    Y
 8  12/12/84  Green,  J L      976-52-3476  DMC       7.22  ppmv   Y    N
```

Figure 2. Sample data as presented in Lotus 1-2-3.

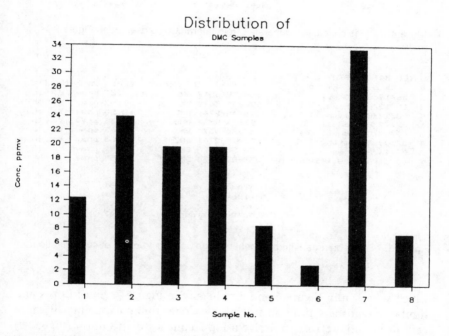

Figure 3. Distribution of DMC samples generated by Lotus 1-2-3.

The sample results may be graphed, which the Lotus program permits (Figure 3). There are also a number of outside programs which will permit even more sophisticated graphs than does Lotus.

Simple statistics are shown in Figure 4. Minimum, maximum, mean, and standard deviations for the ungrouped data were obtained by the use of some very simple commands embedded in the Lotus language. Figure 5 shows the same data but sorted in ascending concentrations. The ability to sort allows the user to do a cumulative frequency distribution

```
SAMPLE DATA - SIMPLE STATISTICS
 #    Date      Employee           ID           Mtrl    Conc   Unit  TWA PPE
 1  02/12/84  Jones,    D F    253-84-9374     DMC     12.35  ppmv   Y   N
 2  02/16/84  Green,    J L    976-52-3476     DMC     23.91  ppmv   Y   N
 3  03/09/84  Jones,    D F    253-84-9374     DMC     19.63  ppmv   Y   N
 4  03/09/84  Smith,    T R    398-45-2741     DMC     19.63  ppmv   Y   N
 5  03/22/84  Smith,    T R    398-45-2741     DMC      8.50  ppmv   Y   N
 6  06/19/84  Jones,    M V    754-28-3519     DMC      2.81  ppmv   Y   N
 7  08/31/84  McGillicudy A O  358-82-1473     DMC     33.54  ppmv   Y   Y
 8  12/12/84  Green,    J L    976-52-3476     DMC      7.22  ppmv   Y   N

                            Minimum              2.81
                            Maximum             33.54
                            Mean                15.95
                            Std. Deviation       9.47
```

Figure 4. Sample statistics generated from information presented in Figure 2.

```
SORTED DATA - SIMPLE STATISTICS
 #    Date      Employee           ID           Mtrl    Conc   Unit  TWA PPE
 6  06/19/84  Jones,    M V    754-28-3519     DMC      2.81  ppmv   Y   N
 8  12/12/84  Green,    J L    976-52-3476     DMC      7.22  ppmv   Y   N
 5  03/22/84  Smith,    T R    398-45-2741     DMC      8.50  ppmv   Y   N
 1  02/12/84  Jones,    D F    253-84-9374     DMC     12.35  ppmv   Y   N
 3  03/09/84  Jones,    D F    253-84-9374     DMC     19.63  ppmv   Y   N
 4  03/09/84  Smith,    T R    398-45-2741     DMC     19.63  ppmv   Y   N
 2  02/16/84  Green,    J L    976-52-3476     DMC     23.91  ppmv   Y   N
 7  08/31/84  McGillicudy A O  358-82-1473     DMC     33.54  ppmv   Y   Y

                            Minimum              2.81
                            Maximum             33.54
                            Mean                15.95
                            Std. Deviation       9.47
```

Figure 5. Same data as Figure 4, generated in ascending order of concentration.

which is shown in Figure 6 and graphically in Figure 7. Since Lotus and similar spreadsheet programs also have logarithmic capability, it takes very little initial effort to derive a log frequency distribution (Figure 8) and curve (Figure 9).

The advanced features programmed into Lotus and in almost any data base program also allow sorting and finding specific results. In Figure 10, the system was asked to find all samples pertaining to a single individual. Note that the criterion is given in the center section, and the names of the individuals are pulled out and printed on the screen.

The system can also be given certain logical commands. For example, the system was asked to find and list all samples that show a concentration of greater than 15 parts per million. Again, this has been pulled out (Figure 11).

Of the flat files available, probably the least expensive and yet, one of

```
SORTED DATA - FREQUENCY DISTRIBUTION
 #   Date      Employee          ID         Mtrl    Conc   Unit  Rank  Cumu %
 6  06/19/84  Jones,  M V    754-28-3519    DMC     2.81   ppmv   1    0.14
 8  12/12/84  Green,  J L    976-52-3476    DMC     7.22   ppmv   2    0.29
 5  03/22/84  Smith,  T R    398-45-2741    DMC     8.50   ppmv   3    0.43
 1  02/12/84  Jones,  D F    253-84-9374    DMC    12.35   ppmv   4    0.57
 3  03/09/84  Jones,  D F    253-84-9374    DMC    19.63   ppmv   5    0.71
 4  03/09/84  Smith,  T R    398-45-2741    DMC    19.63   ppmv   5    0.71
 2  02/16/84  Green,  J L    976-52-3476    DMC    23.91   ppmv   6    0.86
 7  08/31/84  McGillicudy A O 358-82-1473   DMC    33.54   ppmv   7    1.00
```

Figure 6. Frequency distribution generated by Lotus 1-2-3.

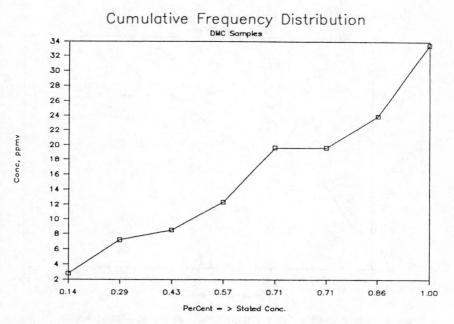

Figure 7. Cumulative frequency distribution graphically generated from information in Figure 6.

the most powerful is a shareware program, PC File III. This program retails for about $50, but a shareware starter can be obtained for approximately $5. This program permits one to add, modify, delete, copy, and move records from one file to another as well as indexing, sorting, and permitting report printouts.

Other relational data base files include dBase III, dBase III Plus, and R:5000.

```
SORTED DATA - LOG FREQUENCY DISTRIBUTION
 #     Date    Employee         ID         Mtrl   Conc   Log  Rank  Cumu %
 6   06/19/84  Jones,  M V   754-28-3519   DMC    2.81   0.44   1    0.14
 8   12/12/84  Green,  J L   976-52-3476   DMC    7.22   0.85   2    0.29
 5   03/22/84  Smith,  T R   398-45-2741   DMC    8.50   0.92   3    0.43
 1   02/12/84  Jones,  D F   253-84-9374   DMC   12.35   1.09   4    0.57
 3   03/09/84  Jones,  D F   253-84-9374   DMC   19.63   1.29   5    0.71
 4   03/09/84  Smith,  T R   398-45-2741   DMC   19.63   1.29   5    0.71
 2   02/16/84  Green,  J L   976-52-3476   DMC   23.91   1.37   6    0.86
 7   08/31/84  McGillicudy A O 358-82-1473 DMC   33.54   1.52   7    1.00
```

Figure 8. Log frequency distribution generated by Lotus 1-2-3.

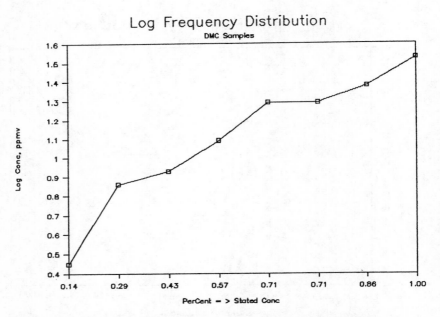

Figure 9. Log frequency distribution graphically generated from information in Figure 8.

Text editing is a mixed bag. In PC-DOS or MS-DOS, there is a function known as EDLIN, a text editor. It is not convenient to use, nor is it simple. Probably the best text editor around is IBM's own Personal Editor II, which is very simple to use. In the latest version, it supports color and split screens at a cost of less than $50.

Word processing is more complex. The leading commercial programs are Word Star, WordPerfect, Microsoft Word, and Multimate. However, all are rather expensive. Even by mail order they will sell for $200 to

```
SAMPLE DATA - FINDING SPECIFIC RESULTS
#    Date     Employee         ID           Mtrl   Conc   Unit  TWA  PPE
1   02/12/84 Jones,   D F    253-84-9374   DMC   12.35  ppmv   Y    N
2   02/16/84 Green,   J L    976-52-3476   DMC   23.91  ppmv   Y    N
3   03/09/84 Jones,   D F    253-84-9374   DMC   19.63  ppmv   Y    N
4   03/09/84 Smith,   T R    398-45-2741   DMC   19.63  ppmv   Y    N
5   03/22/84 Smith,   T R    398-45-2741   DMC    8.50  ppmv   Y    N
6   06/19/84 Jones,   M V    754-28-3519   DMC    2.81  ppmv   Y    N
7   08/31/84 McGillicudy A O 358-82-1473   DMC   33.54  ppmv   Y    Y
8   12/12/84 Green,   J L    976-52-3476   DMC    7.22  ppmv   Y    N

            Employee              Conc
            Green, J L

#    Date     Employee         ID           Mtrl   Conc   Unit  TWA  PPE
2   02/16/84 Green,   J L    976-52-3476   DMC   23.91  ppmv   Y    N
8   12/12/84 Green,   J L    976-52-3476   DMC    7.22  ppmv   Y    N
```

Figure 10. Specific results for one individual generated by Lotus 1-2-3.

```
SAMPLE DATA - FINDING SPECIFIC RESULTS
#    Date     Employee         ID           Mtrl   Conc   Unit
1   02/12/84 Jones,   D F    253-84-9374   DMC   12.35  ppmv
2   02/16/84 Green,   J L    976-52-3476   DMC   23.91  ppmv
3   03/09/84 Jones,   D F    253-84-9374   DMC   19.63  ppmv
4   03/09/84 Smith,   T R    398-45-2741   DMC   19.63  ppmv
5   03/22/84 Smith,   T R    398-45-2741   DMC    8.50  ppmv
6   06/19/84 Jones,   M V    754-28-3519   DMC    2.81  ppmv
7   08/31/84 McGillicudy A O 358-82-1473   DMC   33.54  ppmv
8   12/12/84 Green,   J L    976-52-3476   DMC    7.22  ppmv

            Employee              Conc
                *                +p45>15

#    Date     Employee         ID           Mtrl   Conc   Unit
2   02/16/84 Green,   J L    976-52-3476   DMC   23.91  ppmv
3   03/09/84 Jones,   D F    253-84-9374   DMC   19.63  ppmv
4   03/09/84 Smith,   T R    398-45-2741   DMC   19.63  ppmv
7   08/31/84 McGillicudy A O 358-82-1473   DMC   33.54  ppmv
```

Figure 11. Specific sort for exposures greater than 15 ppm generated by Lotus 1-2-3.

$250. If one wishes to go to the shareware or other inexpensive programs, consider PC Write or Textra. Either can be obtained for approximately $10, although if support and updates are wanted, programs must be registered and the prices go up, but still stay well under $100.

There are many expensive commercial graphics programs available. The PC Picture Graphics (PCPG) program is available through shareware, clubs, or the IBM distribution channel for a very nominal price.

More expensive and versatile graphics and programs include Diagraph, Grafix, and Sign Master.

Just a word about slides. Commercial equipment is available for making slides from the computer programs. A single lens reflex camera with a telephoto lens and macro capability is just as useful. Be warned, however, because of the nature of the signal on the monitor screen, exposure times should be longer rather than shorter.

Communications give one the opportunity and the ability to send data from one location to another or even across the country by telephone line. Most communications use Hayes or Hayes compatible modems. The two most popular software available are the commercial program CrossTalk XVI and PC-Talk (which is available as shareware).

Let me reemphasize that the reference to programs throughout this chapter does not imply that other programs cannot do as well, or even better than the ones indicated. They represent programs with which I am familiar, some of which I have used on occasion, and others which I use regularly. The important part is to determine what it is you need to do. Speak to other computer owners in your area or through the industrial hygiene computer fraternity. It is easy to get into computing. You will find that once you have gotten over the initial hurdle of dealing with computers your productivity will go up immensely.

CHAPTER 4

DEPENDABLE INFORMATION CAN DEFINE THE ENVIRONMENT

RAFAEL H. ESTUPINIAN, PhD and LYNNARD NASH

Controller and President, respectively, Azimuth Technologies, Inc., Pasadena, California

INTRODUCTION

Everybody is concerned about the environment, and everybody is concerned about hazardous substances. The burden of these concerns, however, ultimately falls upon two corporate representatives, the safety professional and the environmentalist.

Because of public awareness, and the complex legislation which has resulted from it, the safety professional and the environmentalist must gather, archive, and manage information which defines, in detail, the various environments with which their corporation or governmental entity is concerned. A clear definition of all aspects of these environments was never more needed, and the need to gather this type of information increases almost daily. The consequences which can result from a lack of organization in the data collection process can be disastrous in terms of: human injury, illness, and even death; the potential for litigation; negative corporate publicity; and the effect upon organizational psychology.

The current focus upon hazardous substances has created even more complex information management requirements. Dependable information about viable solutions for the many problems involved in the development of quality information management systems, which will meet all legislative requirements, is not always readily available. In addition, a busy manager, preoccupied with more pressing concerns, lacks the necessary time to evaluate the information that does exist.

The purpose of this chapter, while not a panacea, is to outline some recommendations and suggestions which may lessen the tasks faced by the corporate safety professional and the environmentalist in the process of evaluating and identifying an information system which will meet his/her particular requirements.

INTERRELATED RESPONSIBILITIES

Even though the safety professional and the environmentalist each have a proper and needed function, both types of managers would agree that their responsibilities often intertwine and that their ultimate objectives can be mutual: *a concern for the individual, and a responsibility to the organization.*

A quick perusal of the types of information which is managed can indicate significant similarities. The safety professional, for example, manages data which can be put into four general categories: safety, health, risk, and training. Each of the broad categories, in turn, will generate a myriad of data collection needs.

Safety, Health, Risk, and Training

Safety, for example, will involve (1) the need to record the occurrence and frequency of occupational injuries and accidents; (2) information to be used in the generation of the OSHA 200 log; (3) the retrieval and updating of illness information, again, for inclusion in OSHA reports; (4) the scheduling of and results of inspections; (5) Workers Compensation information; (6) medical records; (7) the sampling of personnel records; (8) workplace monitoring; and (9) a multitude of overview and statistical reports.

Health information, of course, interrelates with safety information inasmuch as the manager views the very same information, but with another question in mind—Is the corporate environment one which contributes to the poor health of the individual or group?

Risk information, again, is retrieved from the same sources, the medical records, the inspection logs, the Workers Compensation claims that have been paid. Another question is kept in mind— Is the individual, or group, at risk in the corporate environment?

An overview of health, safety, and risk information leads inevitably to training considerations. Has training been appropriate, of sufficient frequency? Has the individual, or the group, been trained in order to be able to exercise appropriate precautionary measures? Has the company met its training responsibility?

CORPORATE REPORT CARDS

Finally, all of the previous questions — Is the corporate environment safe? Is the corporate environment healthy? Is the group or individual at risk in the corporate environment? Has the company met its training responsibility? — can be answered if the information management system is organized in such a way that it can generate detailed statistical reports and overviews. The result is a type of corporate "report card" which can avoid potential litigation, the payment of unnecessary claims, and meet the responsibility for providing a safe, healthy, and risk free environment, and which would be of value to both the safety professional and the environmentalist.

In addition, a safety professional must be able to organize and manage information about hazardous substances and, in turn, respond quickly to the requirements of the recent "Right to Know" legislation.

ENVIRONMENTAL CONSIDERATIONS

The "Right to Know" legislation requires that a corporation, agency, or other entity that produces, utilizes, transports, or otherwise handles a hazardous substance(s) must take steps to ensure that the public and their employees be provided with critical information about that substance(s). It is the right of these target persons to know the essential information about actual and possible hazards.

The primary vehicle for disseminating "Right to Know" information has been a document known as the MSDS form, an acronym for Material Safety Data Sheet. There is a great deal of inconsistency in the generation of the MSDS form. Many manufacturers of hazardous chemicals and substances will produce a material safety data sheet and send it with each chemical or substance that is sent to a customer. However, many of the manufacturers do *not* do this, and the client, or customer, who receives the substance must generate the MSDS form. The production of the form is further complicated because chemicals and substances can be identified under a variety of labels: scientific name, by chemical composition (e.g., H_2SO_4), trade name, synonym, RTEC number, and CAS number. A further complication is that the supporting legislation which may exist can impose specific reporting requirements. These can vary with a particular state, county, or even a municipality.

Defining the Environment

Environment, for purposes of this chapter and for purposes of information management, is defined in terms of the interaction between a specific individual, or group of individuals, and their immediate environs. This information must also include the degree of safety, or peril, which is inherent in the environs. Consequently, the type of information which must be managed is that information which defines this interaction. This way of "looking" at the environment will meet the reporting needs and legislative mandates facing both the safety professional and the environmentalist.

Analogous Data

If we think of the environment in terms of the information which we need about the environment, we can readily see that, in a relatively large corporate structure, both the safety professional and the environmentalist may be collecting analogous data, each with a different focus. (In relatively small and uncomplicated corporate structures, only one scenario may exist because both functions are performed by the same individual.) The data collection scenario for the safety professional and the data collection scenario for the environmentalist, while markedly different, will certainly include areas of similar information. Because of this, greater efficiency, in terms of time and effort, will result from a well planned data management program which produces data which can be shared and which eliminates any duplication of effort that may occur.

Micro Cosmos and Macro Cosmos

In defining the environment in terms of the information needed about the environment, the needed data may be further qualified because the individual, or group, may interact with a smaller workplace environment, or with a larger or more global environment, as would occur during the transport and/or disposal of hazardous substances.

The "workplace" represents one of two "cosmos" within which the environmentalist must collect and manage information. One possible view and distinction might be to think of the workplace as a "micro" cosmos, as opposed to the larger "macro" cosmos of the exterior environment.

Contemporary concerns necessitate a focus upon the hazardous substances which exist in the workplace; their location, their degree of toxicity, the extent to which an individual or group is exposed, the duration of that exposure, etc. Even though the focus for information collection is different from that of the safety or health professional, the answers to the questions asked by the environmentalist can often be found in the same data sources: the employee medical records, the inspection reports, the workplace monitoring reports, or the illness and injury logs. In addition, there must be a concern for training and, of course, a need for statistical and overview reports.

The most common needs for information management and reporting which occur in the "macro" cosmos of the exterior environment, so called because of its almost global nature, are those related to the generation, transporting, and ultimate disposition of hazardous substances (often, "wastes"). Broadly, this information centers upon the three key players in this global scenario: the generator of the hazardous substance, the transporter(s) of the hazardous substance, and the entities involved in the ultimate disposition of the hazardous substance.

If the concern for information management is limited to the smaller and more restricted environment of the workplace, an additional question must be answered, i.e., how long was the individual or group exposed to a particular environment? In other words, was exposure "chronic" or "acute"?

REQUIREMENTS FOR AN INFORMATION SYSTEM

An information management system which would archive information about hazardous materials in the workplace environment must include the following areas: chemicals (substances, compounds, agents, etc.), employee exposure (acute and chronic), the material safety data sheet (MSDS), professional variation opinions, and reports in various formats. This system would need to incorporate multiple data bases which would have the potential or capability for interacting, in order to provide mutually useful information.

In addition, the development of a system to manage this type information must be very subtle and multi-indexed. It must be a flexible information management system which will enable the manager to focus upon exposure either by population or individual. The manager should be left free to determine whether the major effort for data collection will be upon "chronic" or "acute exposure." There must be a provision for recording various types of professional judgments and solutions. The system should be able to generate reports in a format which would be useful for possible epidemiological studies. Information should be managed in such a way that it is readily available for the generation of the MSDS form and other required reports, as shown in Figure 1.

Added features which may be necessary for a complete system are the security of several levels of password for entry, and full data encryption.

A key feature of this system should be *simplicity of use*, even for the relatively unsophisticated microcomputer user. Programs should be menu driven and, in addition to a main menu, incorporate sub menus which make available routines for file merging, index rebuilding, posting, etc.

The program should contain multiple visual and hidden data entry screens which can be used to input and review exposure information. It

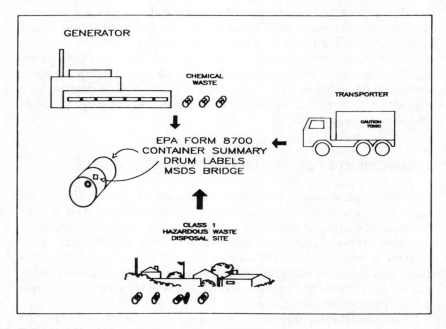

Figure 1. Needs in management of hazardous waste.

should be able to sense a need for additional information based upon program flow, and automatically display or prompt for display any required information.

The company which produces the program should be able to provide or develop various types of conversion routines to enable a client to download from an existing data base.

It must be noted that only vertical software which can be fully customized will fulfill all of these requirements.

A system for managing information related to the exterior environment must be created with three "key players" in mind: (1) the "generator" of the substance, (2) the transporter of the substance, and (3) the destination facility (disposal or repository). It should be generic in format so that it can be customized (modified) to a manager's requirements. Again, such a system must be multi-indexed and have the capability to archive many types of unique information (Figure 2).

Finally, the two information management systems, for the workplace environment and for the exterior environment, must be related in such a manner that they can be used to generate the MSDS document since this would be "essential" to the archiving of information.

However, it should be noted that microcomputer software that merely duplicates the MSDS form is of limited value. Most managers have already arrived at a variety of solutions, e.g., hard copy storage, storage on microfiche, or the use of a word processing program. All seem to agree that archiving and generating the information to produce the MSDS document with a microcomputer is of far more immediate value.

AN OPTIMUM SOLUTION

The optimum solution to meet today's requirements for information management in the safety and environmental field is a microcomputer application which employs "customized" vertical software. In other words, software for a specific market which can be customized to meet the specific requirements of a corporation. The base from which the system is created should be a generic format which is inclusive enough to meet most requirements but "adaptable" to the specific needs of a company.

Such an "ideal" system is in reality a series or group of interrelated programs which can be augmented and from which individual programs can be eliminated if they are of no value to the manager. This series or group should be related in such a way that various data bases can interact to generate the formats which the manager requires, in the format which

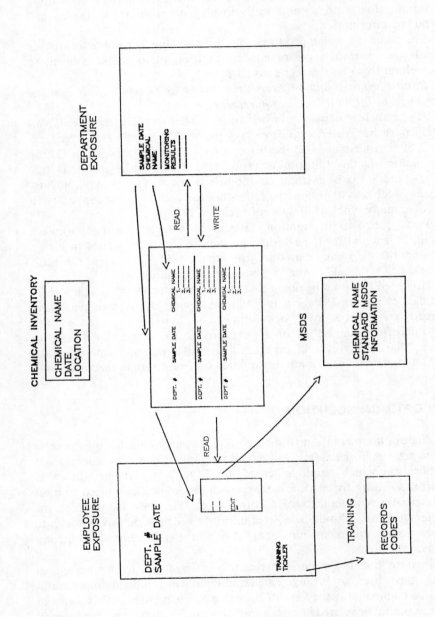

Figure 2. Interacting data bases.

the manager requires. This "optimum" solution is possible in today's technology.

The business microcomputer has "arrived" and coupled with the use of vertical software which can be customized, it offers a solution which was formerly possible only with an expensive mainframe solution. If the software used with the micro can include the capability of archiving and relating multiple data bases, the solution is truly "optimal."

MULTIPLE DATA BASES

What are multiple data bases and how do they function? These data bases are multidimensional. A valid analogy would be with a telescope in which one section opens to another. In the same way, one data base would lead to another data base until the process is complete. For example, in the tracking of hazardous substances, a manager would be able to produce MSDS forms in standard format, keep a record of chemicals and retrieve them by I.D. number, generate hazardous waste manifests, maintain a record of container shipments, etc. Upon the completion of a waste manifest, the system should automatically take you to the element needed and in the proper progression.

Throughout this optimal system, there is a need for variable length note pads in order that a manager be able to store critical information, or even incidental information which might be needed later.

RECOMMENDATION

Increased public awareness and media attention have created greater concern about the environment. As a result, the safety professional and the industrial hygienist face more complex tasks related to the management of information about the environment.

The current legislative focus upon hazardous substances has added to an already burdensome information management task. The busy manager is faced with the task of identifying a viable information management system.

An assessment of the type of information that needs to be managed will result in a determination of the qualities needed in an optimal information management system. In an appraisal of today's technology, the optimal and most cost effective system appears to be a microcomputer solution which utilizes vertical software which can be customized.

The optimal system must have the capacity for multi-indexing of data bases in order to provide a total solution.

CHAPTER 5

AUTOMATION OF MATERIALS INVENTORY AND MSDS INFORMATION

ANDREW J. BECKER

Stewart-Todd Associates, Inc., King of Prussia, Pennsylvania

INTRODUCTION

A software system designed to manage chemical exposure and hazard information can become a powerful tool for an occupational health professional. A system of this type has the potential to tremendously assist in improving the accuracy, completeness, and accessibility of chemical information within an organization. Key data correlations, some of which may be impossible to identify in a manual system, should be made quick and easy. This comprehensive chemical data base can also serve as a resource to individuals with responsibilities such as purchasing, environmental or materials management. However, care must be taken to ensure that the system will operate correctly and that maintenance requirements will not be excessive.

This chapter will cover some of the major considerations that should be addressed in the design or evaluation of automated Inventory and MSDS systems. Chemicals, in the context of this chapter, will be any substance, mixture, physical agent, or other material of occupational health concern. System features to provide data accuracy, ease of entry, retrieval and storage efficiency will be discussed as well as approaches to organizing and maintaining this data.

CHEMICAL/MATERIAL IDENTIFICATION

Critical in establishing and maintaining any chemical information system is the ability to identify the chemical records correctly. For example, in an Inventory/MSDS system, this means having the capability of identifying *all* locations where a chemical in question is used or stored, possibly the quantities of the chemicals at these locations, and all available information on handling and disposal of chemicals or waste material.

However, because chemicals and wastes have multiple naming conventions and, in many cases, coded identifiers created by manufacturers, purchasers, and government agencies, some records for a particular chemical may become disassociated due to naming differences or misspelling. To overcome this problem, the automated system should have a reference function containing a list of valid chemical names or codes. Inventory or MSDS records would be checked against this chemical reference function to ensure the accuracy of any records being created, changed, or deleted. Incorrect spellings or nonlisted chemical names would be flagged for appropriate action by the user of the system.

With a reference function that allows the user to identify chemicals by only a single naming convention, many chemicals flagged as unrecognized may in fact exist in the reference function under another name. For example, methyl ethyl ketone may be flagged as not being on the reference function when in fact a record for 2-butanone exists. The user must make the mental translation to the system's nomenclature and avoid the mistake of adding the new name as a new reference chemical.

A more complex approach to creating the chemical reference function, but one that provides much more flexibility, is to incorporate an alias or synonym function (Table I). This feature allows the users to list alternate naming or coding associated with a particular chemical and will recognize the chemical by a listed alias in data entry and retrieval operations. Here, editing is required to ensure that only unique aliases and chemical names are allowed, i.e., a given alias can be associated with only one chemical. Editing should also ensure that the name selected as the proper chemical name appears only once in the chemical reference function. The user must still be careful not to confuse nonlisted aliases and nonlisted new chemicals. However, the potential for this type of error decreases as the data base becomes more complete. Errors also become more readily apparent than in a system allowing only one naming convention.

Since the object of many data queries is to identify a group of chemicals with certain similarities or characteristics, the chemical reference

TABLE I. Chemical Aliases

Chemical Name	Alias
Ethyl acetate	Acetic acid, ethyl ester
	Acetic ether
	Acetidin
	Ethyl acetate
	Ethyl acetic ester
	Vinegar naphtha
Ethyl acrylate	Acrylic acid, ethyl ester
	Ethyl acrylate
	Ethyl propendate
	Ethyl-2-propendate
Ethyl alcohol	Ethyl alcohol
	Etoh
	Grain alcohol
	Ethanol

function may also contain information to make these relationships possible. Examples of data elements that come under this category are toxicity codes, chemical-family indicators, or codes indicating chemicals with additive health effects. To provide greater accuracy, this data could be checked against reference tables containing values acceptable to the system. Numerical values such as flash point, exposure limits, or reportable quantities may also be useful for identifying chemicals with values above or below a user specified standard.

CHEMICAL/MATERIAL INVENTORIES

A chemical inventory relates chemicals to people and/or places. The two primary editing concerns for inventory purposes are proper identification of the inventory location/personnel group and proper chemical identification.

As mentioned previously, chemical editing is addressed by the system checking inventory entries against the chemical reference function. Editing of location/personnel groups is accomplished by checking against a table or file containing valid inventory codes and their associated descriptions.

Inventory Organization

Typically, the inventory unit will be part of a hierarchical inventory structure where, for example, a company is composed of plants or divisions, a plant has processes or departments, a process may have subprocesses and so on. Developing this inventory structure can be one of the most difficult tasks associated with automating chemical inventory information. It can impact the manner in which data is collected, the level of maintenance the system requires and ultimately, the usefulness of the data collected.

Having the environmental or occupational health chemical inventory follow the lines of existing coding structures, such as building, room, or department codes, offers the advantage of being consistent with other systems and, therefore, more recognizable to users than an entirely new structure. However, the level of structures provided by existing systems may be insufficient or too detailed for some occupational health purposes. Insufficient structure will not provide the information required. (Is it acceptable to identify a chemical as being present in the plant or does the user need to know more precisely where it is stored and who may be exposed?) On the other hand, creating levels of inventory structure beyond that where significant relationships are made will greatly increase the number of inventory records created and, consequently, increase the amount of system maintenance required.

The type of work performed, especially when attempting to identify employee exposures, should also be considered in the design of the inventory structure. In a process/work station-type environment, employing a relatively stationary work force, the inventory structure can be based on physical location codes such as building or room numbers.

However, in an environment where employees are less stationary, physical location-based inventories fall short in their ability to illustrate employees' true exposure to chemicals. In this type of environment, it is best to create inventory units identifying groups of employees. The chemical information collected in this inventory scheme would actually be a chemical exposure profile for a given group of employees.

Additional data collected as part of the inventory record usually relates to the use or state of a material at the inventory unit. For example, paint A is used at inventory 1 in small quantities with low exposure potential. Paint A is used at inventory 2 in large quantities with high exposure potential. Use of paint A and inventory 3 was discontinued November 1984. Examples of data associated with inventory records are shown in Tables II and III.

TABLE II. Inventory Data Element Examples

Inventory unit code: a code for an inventory location — or job classification — department code for establishing an exposure profile.

Material name: name of chemical, product, physical agent, etc.

Start date: the date material was first used in or by the unit.

Stop date: the date material was last used in or by the unit. The system also allows restart of materials showing a break in continuity of usage.

MSDS date: Is a data sheet available at the unit for a given substance? What is the date of the MSDS? Identify MSDS availability at this level to show exactly what information an employee has access to. Multiple versions of the same data sheet may be available throughout a corporation.

Material status code: e.g., W = waste; P = purchased material; I = intermediate product.

Exposure type: I = inhalation; S = skin; B = both.

Quantity indicators: the quantity or quantity range typically found at the unit. Detailed quantity information can potentially increase system maintenance requirements dramatically.

Risk assessment indicators: indicators quantifying/identifying exposure levels, exposure frequencies.

Comments section: comments relating to the use or exposure potential of a material at the inventory unit.

TABLE III. List of Locations by Chemical

Sub-Site	Unit	Local Unit	MSDS Date	Start Date	Stop Date	Phys St	Exposure Type	Risk Factor
Chemical Agent: Benzene								
X1	1413	MEK (with toluene)	8206	8111		L	I	OM
X1	1530	Furfural	8206	8111	8402	L	I	OM OM
X3	01001	8-C Crude unit	7706	8207		L	I	OM 6E
X3	2221	15-2B Olefin saturat	8206	8312		L	I	6M 9E
X3	23135	12A API Separator		8312		L	I	10M 3E
X3	24001	Z-1 Maintenance shop	8206	8211		L	I	4E 8E 3E

Updating the Inventory

It must also be remembered that the inventory requires periodic updating. The scope of the updating processes is based upon the following:

1. The level of inventory structural detail (the number of inventory records).
2. The frequency of material/process/personnel changes.
3. The complexity of the inventory record (the number of data elements collected).

Many wonderfully complex inventory systems have been developed that have proven to be maintenance nightmares. The time required for maintaining the ideal system, both in terms of data collection and data entry, may well exceed the time available. This should always be kept in mind when designing a system and creating update schedules.

If drastic changes have not occurred in inventory structure or content, the follow-up data collection process associated with inventory update may be assisted by or turned over to operational supervisors. A typical scenario is to provide these individuals with the output of the latest inventory and to have them indicate any new materials, discontinued materials, or significant change parameters. If this procedure is constructed properly, it should not create a significant amount of work for the supervisor, and it should provide the additional advantage of creating among supervisors an awareness of chemical use and storage concerns.

MATERIAL SAFETY DATA SHEETS

Having MSDS information organized, automated, and up-to-date can provide powerful chemical information capabilities in addition to extensive capabilities when related to an inventory system for identifying and tracking hazardous materials, exposure and related training, protective equipment, and sampling requirements.

MSDS Management

In addressing the task of computerizing material safety data sheet information, a good starting point is to first create an automated function for the management of MSDS information. Its purpose is to provide a tracking mechanism for all data sheets on file, both current and historical. The advantages for automation are:

1. This function is comparatively easy to put in place, requiring much less effort than entering or maintaining actual MSDSs.
2. It will provide a functional management mechanism even before the actual MSDSs are entered, and may, in fact, preclude the necessity of computerizing the entire MSDS.

The MSDS management function actually is an inventory of vendor supplied materials that is structurally similar to the workplace chemical inventory. Valid chemicals from the chemical reference function are related to valid vendors from a vendor reference function. Additional data elements collected as part of the MSDS management function could include dates of MSDSs on hand, the dates MSDSs have been requested, and other information required for tracking and identification. Typical reports from this function include a listing of all vendors for a particular chemical, all chemicals from a specified vendor (Table IV), and outstanding requests for material safety data sheets (Table V).

Computerizing MSDSs

If an organization's intent is to automate all or segments of the data contained in an MSDS, consideration must be given to the way the data will be maintained. Basically there are two approaches.

1. Enter the information as a text document.
2. Capture the information as values of predefined data fields.

In some cases, it may be advantageous to use both methods for a particular MSDS. The advantages and disadvantages of each approach are most notable when dealing with the narrative information that makes up the bulk of an MSDS.

Utilizing the text method, information can be entered and stored as

TABLE IV. Vendor Chemical List for Acme Marine Paints, Inc., 25 State Street, New York, NY 10004

Chemical	MSDS Requested	Latest MSDS	Dist. Manuf.	Inactive Date
Acetaldehyde	02/01/86		Manuf.	
Acetic acid	02/01/86		Manuf.	
Acetic anhydride	02/01/86		Both	
Acetone	02/01/86	11/83	Both	
Acetone cyanohydrin	01/15/86		Manuf.	
Acetonitrile	02/15/86		Both	
Acetophenone	01/01/86		Both	
Acetyl chloride	02/15/86		Both	
Acetylene			Manuf.	
Acetylene dichloride		10/82	Manuf.	
Acme solvent based coating		12/82	Distr.	
Total chemicals: 11				

TABLE V. Outstanding Requests for MSDS Information
Vendor AB00000001—Acme Marine Paints, Inc., 25 State Street, New York, NY 10004

Chemical	MSDS Requested	Elapsed Time
Acetaldehyde	02/01/86	18
Acetic acid	02/01/86	18
Acetic anhydride	02/01/86	18
Acetone	02/01/86	18
Acetone cyanohydrin	01/15/86	35
Acetonitrile	02/15/86	4
Acetophenone	01/01/86	49
Acetyl chloride	02/15/86	4

received from the MSDS supplier, both in terms of content and possibly MSDS format. There is no requirement to have the MSDS or its narrative statements exist in a standard format, although the user could do this if desired. The main advantage to this method is the relative ease with which MSDS information can be entered.

Disadvantages are primarily in the area of data storage and data retrievability. An MSDS may approach 4000 bytes of information and, although many MSDSs may be shorter, systems without data compression capabilities would require this allocation to be made for all sheets. Any appreciable number of records of this size can quickly add up to significant storage, especially if historical versions of an MSDS are maintained. Also, a single record of 3–4K bytes may be approaching or surpassing the single record limitations of some systems, requiring that the MSDS be split into multiple records. A total free text approach also reduces the ease of searching for commonality among MSDSs due to variations in language and MSDS format from one vendor to the next.

An alternative approach, but one that does have a number of advantages over the text method, is the use of codes corresponding to a list or lists of standard MSDS statements. Data storage is reduced dramatically because the standard statements are stored only once on a table file (Table VI). The actual MSDS record contains only the codes corresponding to the statements, with translations back to narrative statement when the MSDS is reviewed or produced. Data base queries are simplified because the terminology is standardized and retrieval processes are based on the use of codes. Maintenance and updating will be simplified, and the information given to employees is very consistent, reducing the potential for terminology misunderstanding.

However, this method requires the MSDS to be translated into an

TABLE VI. Example MSDS Standard Statements

Section 4—Fire and Explosion

Fire and Explosion Hazards

AA	Oxidizer (contact may ignite combustibles)
BB	Extremely flammable liquid (flash point < 20°F)
CC	Flammable liquid (flash point < 100°F)
DD	Flammable gas
EE	Nonflammable gas
FF	Flammable solid
GG	Combustible (flash point 100° to 200°F)
HH	Can be made to burn (flash point > 200°F)
II	Catches fire if exposed to air
JJ	Dust may form explosive mixtures in air
KK	Class A explosive (DOT)
LL	Class B explosive (DOT)
MM	Class C explosive (DOT)
NN	Noncombustible
99	Not determined/no data

Extinguishing Media

AA	Water
BB	Water fog
CC	Chemical foam
DD	Alcohol foam
EE	Dry chemical powder
FF	Carbon dioxide
GG	Sand/dirt
88	None required
99	Not determined/no data

organization's standard phraseology, creating an additional step in the data entry process. Typically, certain sections of an MSDS are easier to standardize than others. For example, creating a standard list of protective equipment is generally easier than trying to standardize toxicity and first aid. Consequently, it is advantageous to have both text and code capabilities available.

Different organizational needs for automated chemical information will vary both in terms of data content and system complexity. However, the need for accurate and meaningful data is constant among all organizations.

SECTION II

New Products

CHAPTER **6**

THE EVOLUTION OF OCCUPATIONAL HEALTH INFORMATION SYSTEMS— AN HISTORICAL PERSPECTIVE

WANDA RAPPAPORT, PhD

Flow General, Inc., McLean, Virginia

INTRODUCTION

To fully understand the development, evolution, and increasingly widespread use of computerized occupational health information systems (OHSs), one must examine the interaction of relevant technological developments, social pressures, legal requirements, and the structures of organizations in which these systems are used (Figure 1). Each of these factors has been a major influence on the history of OHSs and will continue to influence their evolution in the future. It is insufficient, indeed a misleading oversimplification, to consider only hardware generations or any other single variable as a complete explanation for the rapid growth in the number and diversity of OHSs over the last decade. Their evolution is a product of complex interactions among all of the above factors, which, if analyzed and understood, can help us to predict where we are going and where we should be going with the systems of the future.

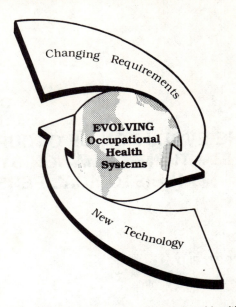

Figure 1. Factors influencing the evolution of occupational health systems.

BACKGROUND

There is considerable variation from organization to organization in the specific details of how OHSs are implemented and used. Fundamentally, however, regardless of organization-specific features, OHSs are information systems for supporting and facilitating two overriding health and safety functions: monitoring the health of employees and monitoring the condition of the workplace. Figure 2 provides a simplified view of the major types of data required to perform these functions within a comprehensive occupational health environment. It also highlights the need for OHSs to interface with other corporate information systems such as those containing personnel and inventory data.

OHSs typically maintain job assignment, health, safety, exposure, training, treatment, and control data so that they can be tracked, related, analyzed, and reported. The ultimate goal is to identify and be able to follow up on possible individual and group problems associated with exposure to hazardous agents in the workplace.

Computerization facilitates the management and analyses of these data for:

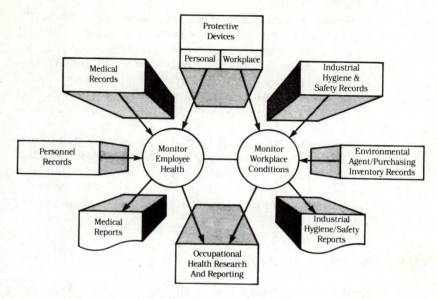

Figure 2. Overview of an occupational health information system.

- *Complying* with federal and state requirements for record keeping and reporting.
- *Reducing risks* to employees' health.
- *Reducing costs* of Workers' Compensation, medical claims, and lost time.
- *Protecting organizations* and their managers from litigation.
- *Facilitating research* in toxicology and epidemiology.

Every company is required by federal and state regulations to report occupational injuries and illnesses on a regular basis. The Occupational Safety and Health Administration (OSHA) has very specific injury and illness reporting requirements. The volume and complexity of the requirements make it extremely difficult to maintain and retrieve this data manually, and even rudimentary surveillance to detect potential health and safety problems is virtually impossible without computerized systems. An increase in storage and reporting requirements is anticipated in the future, as data bases grow and OSHA's new Hazard Communication Rule and often stricter state laws require greater corporate accountability to employees and the community. In a recent survey of FLOW GEMINI users, these considerations were cited as the overriding reasons for automation.

Early detection of adverse health effects or questionably safe conditions not only reduces health risks and costs but also helps avoid legal action. Moreover, in the case of actual or potential litigation, complete, accurate and retrievable data are the best possible defenses for a conscientious organization. Finally, computers permit the rapid access, analysis, and reporting required for special epidemiological and toxicological research as well as more routine surveillance. Indeed, such research is impossible without them.

The three major entities for which data are typically maintained are: (1) employees exposed to hazardous materials, (2) workplaces in which potential health or safety hazards may occur, and (3) environmental agents— physical, chemical, and biological agents that may be health hazards. More detailed data requirements for each of these entities are shown in Table I.

AN HISTORICAL FRAMEWORK

The following historical discussion of OHSs covers only comprehensive systems, those that include sufficient medical, workplace, and environmental agent information to serve the needs of all occupational health professionals. References to specific corporate and commercial systems are included where useful for illustration or reference purposes. The goal is to provide a practical and useful perspective on the evolution of OHSs for those who are intimately involved in utilizing or planning

TABLE I. Data Requirements for a Comprehensive Occupational Health Information System

Employees	Environmental Agents
Work and health history	Physical properties
Medical exams	Ingredients
Laboratory tests	Hygiene
Biological monitoring	Personal protection
Injuries and illnesses	Health hazards
Protection and treatment	Spill procedures
Training	Precautions
Exposures	Transportation
Workplaces	
Sampling data	
Protective devices	
Accidents	
Environmental agents	

for such systems. No attempt is made to include all systems or present a complete historical account of the growth and development of OHSs. Observations and interpretations are based on the personal experience and perspective of the author who has worked with comprehensive health information systems for ten years, first in a research and development setting and during the last five in the development and marketing of FLOW GEMINI, a commercial occupational health information system used by over 30 major corporations in North America and Europe. This experience has provided a unique perspective for understanding the evolution of OHSs.

Technological, Social, and Legal Influences on Early OHSs

The nature of the first OHSs was defined by the state of the art in the computing industry. Systems developed in the early 1970s were of necessity batch-oriented, mainframe systems reflecting available technology. Most corporations already had basic personnel data on a corporate mainframe and a centralized computer staff controlled all computer access and output (Figure 3).

In this environment, Standard Oil Company (Indiana) and Diamond Shamrock Corporation developed the first computerized occupational health systems and were pioneers in defining many of the data requirements listed in Table I. Their systems were developed for internal use. Later both companies marketed their systems which then became the first commercial OHSs.

Existing technology was a factor limiting data input, analysis, and reporting options in these early systems. Although Standard Oil introduced optical scanning, key-entry was the primary data entry mode, and programmers were needed to produce reports. Ad hoc reporting was virtually unavailable.

It is doubtful that the early occupational health systems would have been developed without the impetus of the heightened occupational health and environmental emphasis of the early 1970s. The passage of the Occupational Safety and Health Act of 1970 and the formation of the Occupational Health and Safety Administration, the Environmental Protection Agency, and the National Institute for Occupational Safety and Health, focused corporate attention on the need for greater responsibility and accountability in protecting their employees and the environment. The social responsibility issues that dominated much of the national conscience in the 1960s were now making their mark on corporate responsibility in the 1970s. For the last 15 years, social and legal

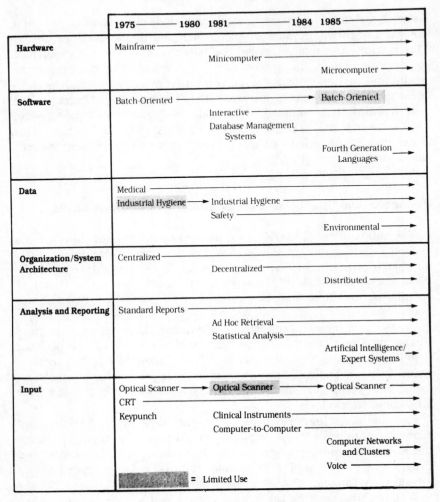

Figure 3. The evolution of occupational health information systems.

responsibility have continued to be a dominant influence on the history of occupational, and more recently, environmental issues in corporate America. Indeed, changing information requirements combined with available technological solutions have continued to determine the evolution of occupational health information systems.

Medical departments and the medical perspective dominated early OHSs and the overall corporate response to occupational health in the

1970s. The newly emerging field of industrial hygiene was little understood by management and physicians, and had little overall corporate influence. Indeed, industrial hygienists themselves were just beginning to understand the complex data storage and analysis problems that were developing. Thus, the 1970s saw the coalescence of complex social, legal, organizational, and technical influences which together provided the stimulus for the development of a new and highly specialized type of information system.

Growing Diversity and Sophistication

Social pressures continued and fostered the further development and refinement of legal requirements. At the same time, management and occupational health professionals became more sophisticated in understanding the problem of health surveillance. It is not surprising therefore that an increasing number of companies were buying and developing computerized OHSs in the early 1980s. At the same time, there were a growing number of technological alternatives in hardware and software, a broader scope of data to be included in systems, many more system architecture and data input options, and growing sophistication in analysis and reporting methods.

The 1980s saw an increasingly widespread use of interactive minicomputers for specialized computing tasks. Large batch-oriented mainframes continued to be the dominant corporate host for centralized administrative functions such as payroll, accounts receivable, and inventory. The minicomputer became increasingly regarded as a natural host for decentralized and/or special-purpose applications. Corporate dissatisfaction with monolithic, centralized administration and control of information combined with the availability of a technological solution in the form of minicomputers led to a growing decentralization of corporate information systems.

Minicomputers also became increasingly widespread in occupational health. They were particularly appealing to occupational health professionals who wanted to control and manage their own unique or sensitive data. Moreover, with the advent of data base management system techniques, users of minicomputers could modify and maintain their own information systems, independent of corporate system staffs who had placed the needs of occupational health professionals so frequently behind those of more central administrative functions or revenue-producing parts of the organization.

Data entry and retrieval were no longer under the exclusive management of corporate systems groups. Minicomputers facilitated decentralized information systems and the management of data input and output by the users—local or divisional medical, industrial hygiene, and safety personnel. In addition, clinical instruments such as audiometers and pulmonary function equipment were being interfaced with minicomputer-based OHSs, eliminating the need to key enter much of the occupational health data. Personnel and inventory data, which previously had been resident only on the corporate mainframe when centralized systems were dominant, could now be selected as needed from the mainframe and downloaded to minicomputer systems. Overall these new systems were characterized by increasingly diverse input and output techniques, including ad hoc retrieval and analysis capabilities. Standard reports developed by programmers were no longer the only option. Indeed, new query languages made virtually any data selection and relationship possible to support special studies and analyses.

The commercial OHS market was now clearly dominated by minicomputer-based systems. There was only one new mainframe system introduced, the Sun Health system. Originally marketed by a subsidiary of its developer, Sun Oil Company, this IBM mainframe system used a general purpose data base management system and offered more flexibility than its mainframe predecessors.

FLOW GEMINI (from Flow General, Inc.), TOHMS (from Tabershaw Occupational Medical Associates), and later DEChealth (from Digital Equipment Corporation) were introduced in the early 1980s. All three systems were comprehensive, minicomputer-based OHSs. It became increasingly clear from their differential success that users wanted a system that they could manage and modify themselves. No matter how comprehensive a commercial software package seemed to be at the time it was developed, some aspects had to be changed to meet the specific needs of users within an organization and to be responsive to new and changing data and reporting requirements. There was also an increasing realization that if data systems specialists or the vendors themselves had to make the changes, users would once more be at the mercy of others. If, instead, users could make their own enhancements, modifications, and additions, the system was much more "theirs." In addition, system maintenance and modifiability ensured users that they were getting what they wanted: a flexible system with growth potential and full user control of schedules and resources.

FLOW GEMINI users repeatedly identify this type of control and flexibility as the most important factor in their selection of the system. The overwhelming majority have tailored the system either with no professional systems support or only temporary, part-time assistance.

It also has been and continues to be important for the vendors themselves to continue to expand, improve, and modify their software products as customers' requirements change and grow in the dynamic occupational health profession. For example, several FLOW GEMINI reports have been modified over time to conform to new OSHA standards and reporting requirements; thousands of MSDSs have been made available to customers in response to the Hazard Communication Rule; and a major Health Physics Module has been developed and integrated for use by customers in the nuclear energy, utility, and pharmaceutical industries.

The 1980s have so far been characterized by an enormous growth in system alternatives. New commercial systems have been made available. Each offers some unique features, and all represent considerably more diversity and flexibility in their data bases, architecture, input, analysis, and reporting options than the early mainframe, batch-oriented systems.

As OHSs have matured and become more widespread, industrial hygiene as a profession has come into its own. Commercial systems contain more industrial hygiene data, and along with safety professionals, industrial hygienists are now full partners in defining system requirements and evaluating, choosing, and using the systems that are developed or purchased. Growing industrial hygiene involvement is evidenced by the three-day ACGIH-hosted symposium on computerized occupational health record systems in the spring of 1983. More than 150 persons attended the symposium and six vendors demonstrated their systems.

Microcomputers, Artificial Intelligence and Environmental Applications

New technologies and changing data requirements created by social and legal pressures continued to stimulate OHS evolution during the mid-1980s.

The most notable hardware-related technological development was the microcomputer. Micros are powerful, relatively easy to master, and have become as widely used and accepted in scientific laboratories as in accounting departments. It was therefore inevitable that microcomputers would become major components in occupational health information systems, both for data input and storage and as delivery vehicles for small or distributed systems.

The mid-1980s offer companies increased options in decentralization. Organizations that have moved from a monolithic management structure and centralized computing to individual plant or divisional control now can choose OHS architectures that permit local and/or regional data bases and data processing.

There appear to be three basic approaches to OHS architecture. Some organizations favor a completely centralized approach to occupational health data management and permit minimal user control. However, that is the exception. Most organizations believe that a single data base assures consistency and compatibility across the company and facilitates epidemiological analysis. Thus they require centralized control over fundamental data base structures, but they want maximum local autonomy in updating and accessing data. A few highly decentralized companies prefer virtually complete local control over data and data base structures with little or no data consolidation, even for archival and analysis purposes. Systems that have the flexibility to respond to any one of these three approaches — centralized, decentralized or distributed — will continue to dominate the OHS market.

As OHSs mature, there is growing attention to the interpretation of increasingly complex and voluminous data. Questions frequently arise about how data can be analyzed to make better and faster decisions to protect the health and safety of workers and the work environment.

This concern coincides with the burgeoning artificial intelligence field and the availability of expert systems. In developing an expert system, occupational physicians, industrial hygienists, epidemiologists, or toxicologists can prospectively define rules for complex decision making and, using expert system "shells" or "builders" that apply one or more reasoning techniques to these rules, develop a system that models a complex decision-making process. Then this expert system can be integrated into their OHS to help interpret their growing data base. As an example, when employees must be screened before assignment to a particularly hazardous or sensitive job, it is frequently important to relate and interpret large amounts of demographic, medical, psychological, and job history data. An expert can predefine decision-making rules using an expert system builder, and the resulting expert system can be exercised from the OHS data base to assist in job placement. This is potentially a much more efficient approach than writing a traditional program or calling on a human expert to examine each employee's record separately. A Flow General subsidiary, General Research Corporation, has developed an expert system builder called TIMM (The Intelligent Machine Model), and the first integration of an expert system into FLOW GEMINI is underway.

The Federal Hazard Communication Rule, state "Right-to-Know" laws, and an increasing number of fatal and near-fatal incidents around the world have led to greater pressure on chemical companies and other organizations to integrate occupational health and environmental information and make it available in a more timely manner to both employees and the public. Air pollution information is the most widely publicized, but many other potential environmental health hazards must be monitored. This requires organizations to collect, store, monitor, analyze, and report environmental data as well as occupational health data. Table II outlines the data required for a comprehensive environmental information system, a major new requirement of the mid-1980s. Figure 4 shows the logical relationship between occupational health and environmental surveillance with their common requirements for hazardous agent reference data and monitoring health and safety problems and events.

It appears inevitable that organizations in the future will foster increasingly closer ties between occupational and environmental health professionals. In response, OHSs will incorporate and integrate environmental data much as FLOW GEMINI did in 1984 when an integrated as well as separate FLOW GEMINI Environmental Information System was announced. System integrations will be facilitated by the proliferation of local area and other networks that can interconnect individuals and by clusters of computers with an increasing array of auxiliary equipment.

New technology and changing system requirements will continue to influence the constantly evolving OHS. The once nearly abandoned optical scanner has found new acceptance and other nonkeyboard data entry modes such as voice and direct sensing instrument interfaces are increas-

TABLE II. Data Requirements for a Comprehensive Environmental Information System

Permits
- Facilities/agencies
- Emission limits
- Renewal requirements

Air and Water
- Schedules
- Emission limits
- Monitoring results

Waste
- Ingredients
- Location
- Generators
- Transportation
- Manifests

Manifests
- Generators
- Transporters
- Disposers

Environmental Agents
- Physical properties
- Ingredients
- Personal protection
- Health hazards
- Spill procedures
- Precautions
- Transportation

PCBS
- Status
- Inspections

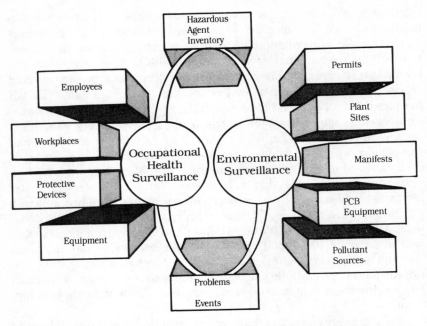

Figure 4. An integrated occupational health and environmental information system.

ing. These alternative data entry modes will be more widely used in the future, providing data base updates in less time and with fewer errors.

The batch systems of the early 1970s have come a long way. It is exciting to ponder the impact that real-time distributed processing, artificial intelligence, and alternative data entry will have on the integrated occupational and environmental information systems of the 1990s.

CHAPTER 7

HEALTH SURVEILLANCE: MANAGING "KNOWN" AND "UNKNOWN" RISK THROUGH INTEGRATION OF INDUSTRIAL HYGIENE AND HEALTH DATA

RICHARD BESSERMAN, MD

Besserman Corporation, SENTRY™ Occupational Health Surveillance Systems, West Glendale Avenue, Phoenix, Arizona

INTRODUCTION

Effective health surveillance requires the cooperation of several disciplines. Its complexity in the workplace has increased in recent years. As new technologies develop, new substances are introduced in the work environment. With experience more is learned about their health effects, the characteristic changes they undergo during a manufacturing process, and their behavior in the finished product. Given the litigious nature of society, government regulations are no longer the only forces influencing the need to implement an effective health management strategy.

KNOWN HEALTH HAZARDS

The fact that a substance is already classified as a known health hazard implies that monitoring guidelines exist and that the agent's toxicology and pathophysiology is fairly well understood. Such information enables members of the occupational health team to develop environmental monitoring strategies and to use such data in the health evaluation process.

The management of "known" health hazards can be simplified by computerization. Automation can assist management in documenting the health and safety of the worksite. Computers are ideally suited to the task of recording monitoring data. They can readily sort and report upon exposure factors, monitoring methods, work processes, worker activities, worker protection, and departmental exposure.

Employee health management requires that the surveillance team establish a method of tracking worker exposure. As one of the developers of the SENTRY™ Occupational Health Surveillance System, the author understands the difficulty one experiences in developing a workable exposure strategy. To simplify the approach, let us first concentrate upon the management of "known" environmental hazards while always accepting the possibility that unrecognized health effects may arise at any time.

A computer's ability to group and sort data is essential in the management of population information. Harnessing this resource enables the system to group employees by exposure. Having estimated relative exposure, it is often possible to divide employee groups by level to enable identification of related adverse health effects.

The approach requires that the environmental engineer identify all "known" exposure factors and their locations. Thereafter, the worksite can be "zoned" to reflect common exposures (stresses). Depending upon the nature of the exposure, it may be possible to subdivide ("zone") a plant by department, work area, work process, job description, or other descriptors that would logically break down the workforce into manageable population groups. The by-products of the effort include the development of a simplified monitoring strategy (Figure 1), the creation of employee exposure histories (Table I), the preparation of material safety data sheets, and the implementation of exposure driven health monitoring (Figure 2).

When a hazard is "known," a system can be implemented to manage risk by monitoring the environment as well as the health of the involved employees. The development of a worker exposure history is important to medical personnel in health monitoring to screen for the most likely adverse health effects.

The process of surveilling "known" risk implies the need for a full understanding of potential health effects. The acute and chronic effects of most "known" hazards are well documented. To simplify surveillance methods and since most agents have organ specificity, SENTRY's developers have divided the employee health records by organ system. This

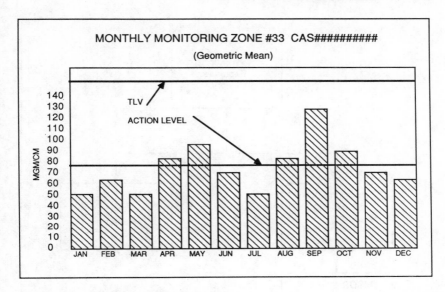

Figure 1. Monthly monitoring for Zone #33.

TABLE I. Chronologic Worker Exposure History

1985/03/29
Employee ID #: Pay Code: Name: Terry K. Francis

Zone	Zone Description	Beginning	Ending	Months
	Work Zone History			
B-02	Chemical testing	1982/08/01	1985/03/29	32
B-14	Heating/cooling	1979/11/01	1982/07/31	33
B-17	Printed circuit assembly	1974/02/01	1975/02/28	15
B-17	Printed circuit assembly	1975/10/01	1979/10/31	49
B-34	Office work	1975/03/01	1975/09/30	6

Factor CAS#	Description	Beginning	Ending	Months
	Exposure History			
C000000202	Glass, fibrous or dust	1979/11/01	1982/07/31	33
C000056235	Carbon tetrachloride	1982/08/01	1985/03/29	32
C000067561	Methanol	1983/03/01	1985/03/29	29
C000071363	Butyl alcohol	1982/08/01	1985/03/29	32
C000071432	Benzene	1982/08/01	1985/03/29	32
P000000020	Heat stress	1979/11/01	1982/07/31	33

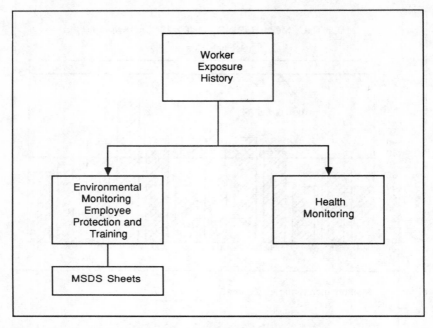

Figure 2. Exposure driven health and safety management.

has led to the successful implementation of a "health effects" module in which health data is aggregated to determine whether abnormalities are exposure related. It is in that context that "zoning" an employee population has enabled our development team to maximize the benefits of computerization. They have built an automated system to compare aggregate data from exposed and comparison populations to detect significant differences in the incidence of adverse health findings.

The computer's inherent ability to identify employee records by user-selected characteristics enables it to subdivide a worker population to uncover other relationships that may have a direct bearing on the detection of a work-related health effect.

PROSPECTIVE HEALTH MANAGEMENT

In a sense, the health management team begins to assume a "prospective" approach to the health and safety of the worksite. The computer becomes much more than a file and record manager. It functions as an analytic tool that is designed to assist the user in managing both "known" and potential health risks.

As health professionals, we recognize the limitations we face in managing the thousands of new substances that have been introduced into the workplace in recent years. We have also noted with great concern new health effects that had not been reported previously. It is increasingly apparent that strategies must be implemented to ensure safety and health. The use of computers will maximize the benefits of gathering information and enable the required analysis. The SENTRY Occupational Health Surveillance System was developed by occupational health specialists with that mission in mind. Its users have already experienced success with the early identification of adverse health effects that were arrested before the development of an occupational illness.

UNKNOWN HEALTH RISKS

If everything were already "known," the complexity of health surveillance would be greatly reduced. Unfortunately, even in dealing with the "known," we make new discoveries that further complicate occupational health management. It is really the "unknown" that requires the implementation of a workable strategy to protect the employee, the company, and even the consumer. Since information is often limited when new substances are added to the work environment, one must be prepared to encounter unexpected problems.

The use of a "prospective" system of surveillance such as that developed in SENTRY enables the user to identify exposure related adverse health effects. By searching for patterns consistent with the prodromal stages of a disease, it is often possible to detect subtle changes early enough to intercede. When successful, the benefits can be beneficial to both worker and employer alike.

The task of managing "prospectively" requires more than a computerized record keeping system. Prospective health management is dependent upon the implementation of a software solution that provides built-in strategies to enable health personnel to identify occupational health problems at an earliest possible stage. The computer is really an extension of the user's mind, expanding its horizon and enabling further exploration and evaluation. It represents a vital resource to all members of the occupational health team, and above all, represents a means of managing risk in the workplace.

For example, a prospective approach to managing risk associated with exposure to an organic solvent might include periodic liver function studies (Figure 3). In addition, a short questionnaire could be included to determine whether the exposed employee group had higher levels of liver enzymes signaling hepatocellular (liver cell) damage or a more specific health complaint. In either event, when an unexplained difference is identified, similar ad hoc surveillance tools are provided to compare worker groups to detect other differences, especially those noted between exposed and comparison (control) populations.

In addition to the ability to study individual historical, examination, and laboratory data, the SENTRY Occupational Health Surveillance System includes a built-in diagnostic module that allows retrospective analysis of nonoccupational disease diagnoses, nonoccupational tumors, and nonoccupational immunologic disorders. It is in this area that careful review of aggregate data may highlight instances where occupational diseases have been mistakenly classified as nonoccupational (Figure 4).

CONCLUSION

Whether the strategy you develop is prospective or retrospective, your success will depend upon a mutual understanding of the needs of each

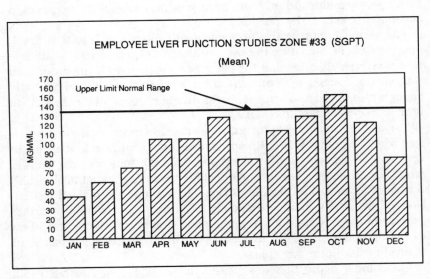

Figure 3. Monthly monitoring of suspected liver dysfunction for Zone #33.

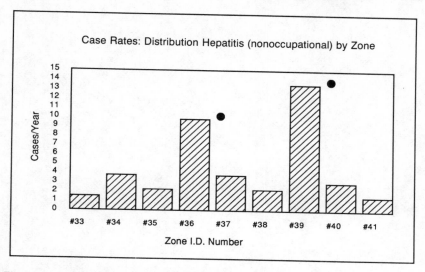

Figure 4. Hepatitis case rates by Zones #33–41 (1985).

discipline. The industrial hygienist or environmental engineer must provide the necessary exposure data to make surveillance effective.

A methodology can be implemented to safeguard the worker, document the health and safety of the worksite, and serve as a steward for a company's products.

SECTION III

Practical Uses

CHAPTER 8

ELECTRONIC SPREADSHEETS IN INDUSTRIAL HYGIENE MANAGEMENT

MITCHELL S. BERGNER, CIH
Honeywell Defense Systems, Minnetonka, Minnesota

WHAT IS A SPREADSHEET?

Accountants and business planners have used various forms of ledgers as their standard tool. The ledger allowed them to arrange quantities of numbers in rows and columns. The numbers could then be added, subtracted, or compared with one another for financial management tasks. With the advent of the personal computer, the accounts' ledger became available as spreadsheet programs marketed under a variety of trade names for all disk operating systems. The software electronically places numbers in the computer's memory in rows and columns and allows for an unlimited number of calculations, comparisons, and "what if" inquiries.

THE BIRTH OF THE ELECTRONIC SPREADSHEET

The electronic spreadsheet began at the Harvard Business School when Daniel Bricklin and Robert Frankston observed their professor going back and forth to the blackboard to calculate tables of numbers only to erase the work and recalculate it as one or more variables changed. They developed a program for the Apple II computer that would recalculate mathematical functions instantaneously as one or more variables changed. This program became known as VisiCalc.

Within a year, VisiCalc had changed the world of microcomputers. It had altered the idea of the small computer from a hobbyist's toy into a useful business tool. By the end of 1982 Bricklin and Frankston's company was netting $7 million per year and was growing at a rate of 100 per cent per year. VisiCorp, which marketed VisiCalc almost from the beginning, had grown to a $35 million per year company. VisiCalc was running on more than half a million computers, including those made by Apple, Hewlett Packard, IBM, Radio Shack, and Atari.

Spreadsheet software has continued to advance by integrating graphics, data base management, and word processing into one program. Anyone managing numbers can benefit by the use of spreadsheets. In industrial hygiene, spreadsheets make it easier to perform calculations and observe trends in data.

Spreadsheets may also be used to word process and track items such as annual objectives and department budget (Figure 1).

APPLICATIONS IN INDUSTRIAL HYGIENE MANAGEMENT

Spreadsheet programs cost about $100 to $500, but their primary function is essentially the same. One can create a grid of values and specify mathematical relationships within the grid. The grid may contain as many as two million locations or "cells" depending on the program and the memory capacity of the computer. The relationships, or formula, are tied to the locations of the numbers, not the numbers themselves, so that a number can change within the grid without upsetting the formula. More expensive programs include capabilities for sorting, graphing, and customizing features that present more calculating and analysis options.

AIR SAMPLING

Suppose, for example, there was a work area that required a number of weekly area samplings. These data could be entered on the grid weekly to calculate time-weighted average, monthly average, and standard deviations. The data could then be plotted on bar, line, or pie graphs, and printed on paper using a dot matrix, laser, or ink jet printer for quarterly or annual reports to management (Figures 2 and 3).

PAGE 1 OF 3   ** INDUSTRIAL HYGIENE OBJECTIVES 1986 **

14-JAN-86   JAN FEB MARCH APRIL MAY JUNE JULY AUG SEPT OCT NOV DEC   TARGET

*** AUDITS ***
    NB103   0!   0!   0!   0!   !QUARTER
    NB502   0!   0!   0!   0!   !QUARTER
    NB104   0!   0!   0!   0!   !QUARTER
    HOPKINS   0!   !QUARTER
    HOPG   !AS NEEDED
    JAAP   !AS NEEDED
    HORSHAM   !AS NEEDED

*** WORKER RTK ***
    TRAINING
    NB502   0!   !
    NB103   0!   ! 2/86
    NB104   0!   ! 3/86
    HOPG   0!   ! 3/86
    SLP   0!   ! 8/86
                                                                                          ! 1/86
*** MISC ***
    HAND TOOL STUDY NB103   0!   !
    ERGONOMICS DEP.SAF.COMM.   ! 6/86
    HAZARDOUS MAT.LABELING   !AS NEEDED
    PE/QE IH COURSE   0!   ! 9/86
    VENTILATION TESTING   ! 8/86
    COMPUTERIZE LOST TIME DATA   0!   !ONGOING
    REPORT INDICES MONTHLY   ! 6/86
                                                                                          !ONGOING
*** CALIBRATION ***
    BENDIX HIGH VOL.   0!   !
    HOLADAY MICROWAVE MTR   ! 1/86
    ALNOR VELOMETER   !DUE 1/88
    QUEST SLM KIT   !
    KURZ MASS FLOW   0!   ! 9/86
    KURZ HOT WIRE   0!   ! 6/86
    GENRAD SLM KIT DONE!   ! 1/86
    METROSONICS DOSIMETER   0!   !4/85 ! 4/86 $250
    AUDIOMETER NB103   0!   !9/85 ! 5/86
                NB502   0!   ! 4/86
                                                                                     !4/85 ! 4/86
*** AIR SAMPLING *** $350
    METHYLENE CHLORIDE (GATOR) 0!   ! 3/86
    DU   (502)   !CHEM LAB
    V204   (SLP)   0!   ! 6/86
    MDA   (103)   0!   0!   ! SEMI
    MICROWAVES (TCAAP)   !NEXT DUE 4/88   ! 4/88 18 LOC

*** NOISE ***
    SURVEY
    NB103   !
    NB502   !AS NEEDED
    NB104   !AS NEEDED
                                                                                          !AS NEEDED
    TRAINING
    NB103   0!   !
    NB502   0!   ! 8/86
    HOPG   0!   ! 3/86
    NB104   0!   ! 7/86
                                                                                          ! 8/86
```

**Figure 1.** A nonmathematical use of VisiCalc. Department objectives are listed and tracked by maintaining a spreadsheet list.
```

MICROCOMPUTER APPLICATIONS IN OH&S

JANUARY 85

DATE	1A	2A	3A	4A	5A	6A	7A	8A	9A	10A	11A	12A
WEEK OF 1/2/85	7.2	12.4	5.2	2.3	14.9	2.8	1	.5	.5	2.4	7.6	2.4
1/8	4.5	10	9.43	3.7	17.1	3.73	.83	.624	2.05	2.26	5.46	6.4
1/14	8.59	3.51	11.6	6.43	23.7	8.06	1.41	.332	1.04	5.84	12.9	4.69
1/22	10.1	3.76	4.63	3.68	15.4	4.15	.318	.119	2.48	2.59	8.3	9.57
1/28	12.2	3.65	1.84	5.58	25.2	2.05	.155	.089	.577	.309	7.79	3.66
AVERAGE	8.52	6.66	6.54	4.34	19.26	4.16	.74	.33	1.33	2.68	8.41	5.34

FEBRUARY 85

DATE	1A	2A	3A	4A	5A	6A	7A	8A	9A	10A	11A	12A
2/4	13.8	2.36	6.52	7.41	34.2	2.33	.234	.117	2.1	4.87	7.25	5.13
2/12	8.65	2.67	9.55	8.13	21.6	1.82	.241	.18	3.91	2.84	3.82	2.85
2/19	5.98	2.13	17.8	3.34	48.2	1.3	.327	.447	3.12	2.26	8.59	5.1
2/26	7.1	2.3	7.1	4.2	21.9	1.3	.79	.05	4.7	1.8	11.5	.1
AVERAGE	8.88	2.37	10.24	5.77	31.48	1.69	.4	.2	3.46	2.94	7.79	3.3

MARCH 85

DATE	1A	2A	3A	4A	5A	6A	7A	8A	9A	10A	11A	12A
3/5	2.67	27.5	5.62	3.04	14.6	1.01	.33	.15	2.8	2.54	2.03	.87
3/12	7.29	1.03	11.5	3.28	9.46	1.21	.49	.38	1.2	3.6	7.19	.96
3/20	3.05	4.35	6.84	4.56	12.5	1.57	.96	.25	5.87	13.3	13.1	4.34
3/26	6.35	2.89	18.2	13.6	31.7	1.09	.76	.63	1.57	4.31	9.21	3.01
AVERAGE	4.84	8.94	10.54	6.12	17.07	1.22	.64	.35	2.86	5.94	7.88	2.3

** DEPLETED URANIUM CONTINUOUS ROOM AIR SAMPLES, 1985 **

```
1A-W.SCRAP WING          7A-MID. HEAT TREAT
2A-E.SCRAP               8A-E. HEAT TREAT
3A-N.DU MFG. WING        9A-N. EAST WING
4A-N.DU MFG. WING        10A-DOCK,EAST WING
5A-MID. DU MFG.WING      11A-W. EAST WING
6A-S.DU MFG. WING        12A-N.WALL EVAP. ROOM
ALL WEEKLY RESULTS REPORTED IN MICROGRAMS OF DU PER CUBIC
METER OF AIR. HONEYWELL ACTION LEVEL = 20 UG/M3
```

Figure 2. Continuous environmental samples at 12 different locations generate voluminous data to be reviewed and stored. A simple spreadsheet can assist in organization of 624 data points per year.

Employee air sampling could be tracked by the operation or by employee name with the time-weighted average calculated by the program. The sheet is set up with the standard formula for time-weighted averages, leaving room for the employee's name, date, and other required information. One cell could be set up to indicate "o.k." or "excessive" by automatically comparing a calculated time-weighted average with the permissible exposure level.

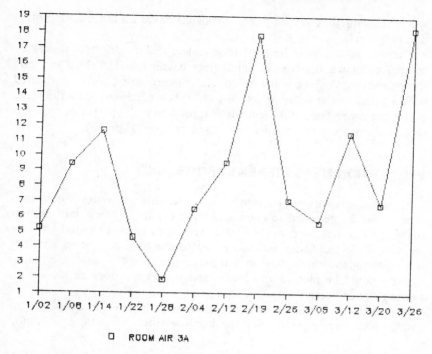

Figure 3. A more advanced spreadsheet, such as Symphony, provides a graphics function for any specified data point. This graph is a plot of the 13 data points from continuous sample 3A, listed in Figure 2.

NOISE SAMPLING

Noise survey data could be entered and calculated into eight-hour exposures. "What if" analyses will assist in administrative control by changing exposure time on the grid to see how long it would take to reach 100 percent of the permissible dose. Changing the data on the sheet results in instantaneous recalculation allowing unlimited experimentation with dose and time. Ear plug attenuation calculations are equally easy to do once the sheet is set up with the proper formula.

Some spreadsheet programs allow you to set up a table of "look up" values used to perform other calculations. For example, a table of attenuation data by octave band for a hearing protector is placed in a "look up" table. When performing attenuation studies, by entering the octave band readings the format will automatically "look up" the attenuation from a table, calculate the appropriate correction, and deduce the average noise reduction and employee exposure based on the use of protection (Figure 4).

The spreadsheet may be used to calculate audiometric test results by comparing annual input to some predefined baseline. OSHA shifts and state compensation calculations for mono- and binaural loss may be automated with a minimum amount of setup time. When combined with the word processing capabilities of the spreadsheet, professional reports can be generated to the medical community or to the employee (Figure 5).

VENTILATION MEASUREMENT PROGRAMS

To manage a survey program on many ventilation systems a spreadsheet could be designed to enter and store original system data. Subsequent evaluation, such as face velocity measurements, would then be calculated electronically and compared to the original system parameters. If the spreadsheet program has graphic capabilities, malfunctioning systems could be plotted and incorporated into a report to plant engineering for corrective action.

Managing increasingly complex industrial hygiene programs is no small task. The responsibility for keeping track of data, statistically

```
         EMPLOYEE PROTECTION UTILIZING "EAR" BRAND PLUGS


OCTAVE BAND         125   250   500  1000  2000  4000  8000      TOTAL
----------------------------------------------------------------------
OB LEVEL ENTER=     100   102   106   110   113   114   104      118. dB
ADJUST FOR dBA      -16    -9    -3     0     1     1    -1
OCTAVE dBA           84    93   103   110   114   115   103      119. dBA
ATTENUATION    =     35    36    41    41    38    44    42
2X STD. DEV.   =    4.4   4.8   5.6   5.8   6.2     9   8.6
dBA EXPOSURE       53.4  61.8  67.6  74.8  82.2    80  69.6      85.0 dBA
----------------------------------------------------------------------

      PROTECTED dBA EXPOSURE    ----> 85.0


      CALCULATED dBA REDUCTION  ----> 33.6
```

Figure 4. Look up tables are used to store constant data for later retrieval. The attenuation data for EAR brand protectors has been entered on this table along with other commonly used devices. By inserting EAR between the quotes in the title, the program searched for the appropriate data and inserted it into the calculations. This routine performs over 30 calculations.

MITCHELL S. BERGNER   119

```
RICHTER REINHOLD  486-28-5820   502
                  10/24/24
                         LEFT EAR                          RIGHT EAR             SHIFT AVERAGES   LOSS TOTALS  CLASS
FREQUENCY       500 1000 2000 3000 4000 6000 8000   500 1000 2000 3000 4000 6000 8000   LE   RE      LE%  RE%  %
                                                                                        A234 A234   MONO MONO BINU
*BASELINE*
10/21/74 -----  35   25   15   50   35   25   X     25   20   15   55   55   60   X                 9.38 5.63 6.25    3
---------
6/17/76  -----  35   30   35   50   55   40   X     20   25   30   55   65   60   X                 18.8 11.3 12.5    3
YR TO BASE       0   -5  -20    0  -20  -15          5   -5  -15    0  -10    0         -13. -8.3
---------
12/12/78 -----  25   25   25   45   25   35   X     30   30   30   55   55   55   X                 7.5  16.9 9.06    3
YR TO YR        10    5   10    5   30    5        -10   -5    0    0   10    5          15  3.33
YR TO BASE      10    0  -10    5   10  -10         -5  -10  -15    0    0    5         1.67   -5
---------
11/10/80 -----  35   35   30   55   55   55   X     25   25   30   65   65   70   X                 20.6 16.9 17.5    3
YR TO YR       -10  -10   -5  -10  -30  -20          5    5    0  -10  -10  -15         -15  -6.7
YR TO BASE       0  -10  -15   -5  -20  -30          0   -5  -15  -10  -10  -10         -13. -12.
---------
6/22/81  -----  30   30   25   60   65   50   X     25   25   25   60   65   65   X                 16.9 13.1 13.8    3
YR TO YR         5    5    5   -5  -10    5          0    0    5    5    0    5         -3.3 3.33
YR TO BASE       5   -5  -10  -10  -30  -25          0   -5  -10   -5  -10   -5         -17. -8.3
---------
6/14/82  -----  30   30   35   55   55   45   X     35   35   30   55   65   65   X                 18.8 20.6 19.1    3
YR TO YR         0    0  -10    5   10    5        -10  -10   -5    5    0    0         1.67    0
YR TO BASE       5   -5  -20   -5  -20  -20        -10  -15  -15    0  -10   -5         -15  -8.3
---------
7/21/83  -----  30   30   30   60   70   50   X     25   25   25   60   60   70   X                 18.8 13.1 14.1    3
YR TO YR         0    0    5   -5  -15   -5         10   10    5   -5    5   -5          -5 1.67
YR TO BASE       5   -5  -15  -10  -35  -25          0   -5  -10   -5   -5  -10         -20  -6.7
---------
6/1/84   -----  15   25   25   55   50   45   20    15   20   25   60   55   70   50                7.5  7.5  7.5     3
YR TO YR        15    5    5    5   20    5         10    5    0    0    5    0          10 1.67
YR TO BASE      20    0  -10   -5  -15  -20         10    0  -10   -5    0  -10         -10   -5
---------
3/14/85  -----  20   25   35   50   60   40   25    20   25   30   60   55   70   55                11.3 13.1 11.6    3
YR TO YR        -5    0  -10    5  -10    5         -5   -5   -5    0    0    0          -5 -1.7
YR TO BASE      15    0  -20    0  -25  -15          5   -5  -15   -5    0  -10         -15 -6.7
```

Figure 5. An audiometric testing spreadsheet. Annual data are entered on successive lines for complete record storage. The program compares annual tests to baseline and the previous test. OSHA shifts and state compensation calculations are performed.

evaluating the data, and observing early trends has always been within the realm of the industrial hygiene professional.

Spreadsheets are well equipped to help industrial hygienists meet one of their primary responsibilities: accurate evaluation and dissemination of exposure and test data. Any application which requires the manipulation of numbers or the maintenance of a list(s) is particularly well suited for today's spreadsheet software. With a little experience using spreadsheets, multitudes of new applications will be evident that will benefit the overall effectiveness of industrial hygiene programs.

CHAPTER 9

DATA BASE MANAGEMENT

DENESE A. DEEDS, CIH

Industrial Health and Safety Consultants, Inc., Huntington, Connecticut

INTRODUCTION

A Data Management Program or Data Base Program is a program designed to store information and allow that information to be retrieved, sorted, searched, etc., in a variety of ways. This ability to allow varied retrieval is the reason these types of programs have become so popular for safety and health applications. A simple Respirator Data Base application will be used for example purposes, but similar data problems that this approach may simplify should be kept in mind. Using this type of program, information on each employee's respirator use can be stored, e.g., name and demographic information, the type of respirator used, medical approval, and training. This data then can be listed or reported back, sorted on the worker's date of medical approval, training date, etc. The usefulness of this type of reporting is obvious. It can speed up scheduling, reduce typing chores, and reduce paperwork by permitting a "file" to be kept once and merely accessed in different ways. At present, records such as these may be kept manually by employee as well as by month of training for scheduling annual training and still another list by date for medical approval review. Using a computer the one file can be quickly, almost instantaneously searched for the data needed today.

Some other useful applications of Data Management Programs include:

1. Maintenance of employee safety information including training received, personal protective equipment issued, accidents, medical visits and complaints, etc.
2. MSDS information.
3. Industrial hygiene sampling data.
4. Mailing lists for newsletters, company reports, etc.
5. Biological monitoring data.

There are generally limitations on the size of data records for any commercially available programs, therefore, consideration needs to be given to limiting the scope of these applications accordingly. One of the strengths of dBase III is that you can create "relational" data bases and keep portions of records, all related by employee Social Security number on separate records. In the menu driven Advanced DB Master, this is not possible.

Some general terms need to be defined before continuing. A "file" is the term used for the format for a particular application. The "record" is one set of entries whose primary key field is unique. A "field" is an individual piece of data. These terms generally describe the structure of a data base. Once the record format is set up, data is entered into the fields. The regurgitation of that data is called a "report."

At this point we will create a respirator data base using Advanced DB Master. The first step in the process is to define what fields need to be included in the record. For this application, these will be:

Employee Name	SS#
Job Description	Plant Address
Type of Exposure	Category < 10 PEL
	10–100 PEL
	> 100 PEL
Respirator Model	Size
Filter Type	
Fit-Test Date	
Medical Approval Date	
Training Date	

We next boot the program and select the create file option. We then enter each field and select the field type and length in response to the program prompts and place the field on the screen. This program allows up to 9 screens for the record and up to 3000 characters for data entry. The first field or fields are the primary key, which must be unique for each record. If we are going to use Social Security number, that can be all or part of the primary key. In this program, the records are maintained in alphanumeric order, so if you want most reports in alphabetical order, a name field should be the first primary key entry (Figure 1).

Once all the fields have been entered, you tell the program that you are finished, and the program directs you to insert a diskette to be formatted.

If we were performing this with dBase III, we would go into the CREATE structure by entering that command or through ASSIST. A DOS formatted disk is needed on which to write the file when completed.

We are now ready to add records. In Advanced DB Master, there are many menu selections relating to the status of the program and functions available. Time does not permit an explanation of all of these, but there are a wide variety of functions built into this program to make it easier to use.

dBase III is a very powerful program and some books actually refer to it as a programming language. dBase can perform all of the functions that are available as menu selections on this program, either by using commands or by programming menus and loops to make it usable for novices. It is a complicated process and one must consider whether the versatility available is worth the effort. One option is to have the program designed for you by a dBase programmer and the menus and reports created for you. If you decide to take on the job yourself, there are some excellent books available to help.

The real power of data management is getting the information out in the way you want. There are two ways to view your data, on the screen

```
Record Length: 253/3000          / Add Field .                    Page 1 of 1
NAME LAST,FIRST!_____

SS# 000-00-0000                  PLANT ADDRESS _____

JOB _____

EXPOSURE _____

CATEGORY:    <10 x PEL _         RESPIRATOR:    MODEL _____
             10-100 x PEL _                     SIZE _____
             >100 x PEL _                       FILTER TYPE _____


FIT-TEST DATE MM-DD-YY           MEDICAL APPROVAL: MM-DD-YY

TRAINING DATE: MM-DD-YY


1 Alphanumeric 2 Numeric 3 Dollar 4 Y/N 5 Date/Time 6 User-Defined 7 Label-Only
  Field Type _      Field Label _____      Display Mode 2
              ^Done with page     ^Edit mode      ^Field Options
```

Figure 1. Example of employee record.

using DISPLAY and BROWSE features, or in the generation of hard copy reports.

First we will create a report from our respirator data base. This will be a simple report listing all employees alphabetically and the type and size of respirator they use. We select the menu selection for creating or modifying reports and find several subformats and a master format selection (Figure 2). These subformats allow mixing to create many standard reports while doing the work of defining the page layout, for example, only once. We will use the standard page layout. Now we will go to the Data Subformat and create the format for this report. We will use no heading but will place column titles of Employee Name, Respirator Type, and Size. Then in the data line section we will place the field numbers for these fields. When finished, the subformat is saved (Figure 3). The subformats can be put together to create a master format and will be called Respirator Selection. Now this report can be printed (Figure 4).

```
/ Build/Edit Report Formats for RESPIRATR.01 .

     Build/Edit a Report Subformat

        ZDDDDDDDDDDDDDDDDDDDD?
        3 1    Page              3
        3                         3
        3 2    Data              3
        3                         3
        3 3    Sort              3
        3                         3
        3 4    Selection         3
        3                         3
        3 5    Printer Control   3
        @DDDDDDDDDDDDDDDDDDDDY

     Or Build/Edit a Master Report Format

        ZDDDDDDDDDDDDDDDDDDDD?
        3 6    Master            3
        @DDDDDDDDDDDDDDDDDDDDY

        Enter your choice (1 to 6):  _
        ^C Main Menu
```

Figure 2. Menu example for creating or modifying reports.

```
                            / Edit Column Titles .
Line        1         2         3         4         5         6         7
 #  12345678901234567890123456789012345678901234567890123456789012345678901234
  1             EMPLOYEE NAME                    RESPIRATOR MODEL      SIZE
    ------------------------------ Print Lines --------------------------------
        NAME LAST,FIRST_____.MODEL_____.SIZE_____




                       ^Edit this line      ^C Data Menu
            ^Justify this line to left  ^P center this line     change line
```

Figure 3. Example of subformat.

```
14-Jul-86              EMPLOYEE RESPIRATOR SELECTION              Page 1
                  EMPLOYEE NAME                  RESPIRATOR MODEL     SIZE
FRANKE, HELEN                               3M PWAP                   STD
SMITH, LARRY                                AO HALF-MASK              LARGE
TAFT, HARRY                                 WILSON HALF-MASK          MED
```

Figure 4. Example of respirator selection report.

```
                        / Insert Mode for Text Block #501 .

^F1^
^F4^
DEAR ^F1^:

YOU ARE SCHEDULED FOR ANNUAL RESPIRATOR TRAINING ON MAY 14, 1986 AT 9:00 AM IN
THE PERSONNEL CONFERENCE ROOM.
PLEASE NOTIFY THE SAFETY OFFICE IF YOU CANNOT ATTEND.


JOE SMITH
SAFETY DIRECTOR

DDDDDDDDDDDDDDDDDDDDDDDDDDDDDDDDDDDDDDDDDDDDDDDDDDDDDDDDDDDDDDDDDDDDDDDDDDDD
       remove last insertion    Escape without inserting   Ins finish insertion
                 ^Field list     ^Printer controls    ^Text block controls
```

Figure 5. Example of form letter using field variables.

Another nice feature offered by this program, and available for most other Data Base Programs using a mail-merge type program, is the creation of form letters using the field variables. A training notice can be created very easily using this facility. The form letter is created as a text block, using the "F" fields where the employee name (F1) and plant address (F4) should appear on the form letter. The computer will print the letters, ready to mail to the affected employees (Figure 5).

Some comments on data entry—remember that the computer sees character strings only when it looks at words—it cannot understand meaning. When data is entered into fields, it must be completely consistent if good retrieval and reports are desired. Placing a hyphen instead of a space, e.g., Dept J-1 and Dept J 1, at different times will mean the computer will see these as different entries and will not combine these records into the same department when executing a department search. Different combinations of upper and lower case can cause the same type of problems. The standard is all upper case data entry to eliminate that source of error.

A data base application in an MSDS program includes the use of the DISPLAY and BROWSE features. In this application, the setup is with the component fields connected in an "array" so that all fields can be searched with one search command. All records in the file can be searched for Sodium Hydroxide to see which materials contain this component. Using the DISPLAY feature the entire record can be reviewed. If the BROWSE function is used, the operator can "flip" through all the records and see only the fields that have been designated. dBase III has BROWSE.

Most Data Base Management Programs also allow some restructuring or change in the record format of the file even after data has been entered. It is wise to familiarize yourself with any restructuring limitations of the particular program you are using. If, for example, you cannot change field lengths once data is entered, you may want to add a few extra characters when you set up the file.

In summary, Data Management Programs provide a very powerful tool in handling industrial hygiene and safety data. Determining when to "computerize" your data can be the hard part. In general, it is probably not worthwhile to set up a program for files rarely used, unless the need to manipulate the data is great. However, once the data is entered, you will find uses for it you never anticipated.

CHAPTER **10**

WORD PROCESSING APPLICATIONS IN INDUSTRIAL HYGIENE

LEONARD WILCOX

Industrial Hygienist, CAL/OSHA Fresno District Office, Fresno, California

Learning to use a word processor is an investment of time and money. Until the user completes the transition from a typewriter or pen and paper, an exorbitant amount of time is spent on the wrong side of the learning curve. The question, then, is: Is the payoff worth the price?

I firmly believe that, for industrial hygienists and any other professional who must communicate clearly, the time spent learning to use a word processor will bring many dividends in the form of improved communications and reduced time spent writing.

Writing is my avocation. When I bought my first word processing system, it was justified by a few sales of articles to magazines. It was an agonizing decision because of the incredible number of systems available at prices much higher than they are today. Yet, I felt that, with the way things were going, the system would pay for itself in a year and a half.

That estimate was beaten by a year. Within six months, my writing quality and output had increased dramatically because of my word processor. When I began using my word processor for my industrial hygiene reports, the amount of time spent writing reports was cut by one-third, and my reports communicated more clearly what I had to say.

How does a word processor do all this? It simplifies the mechanical aspects of writing.

Writing is rewriting. In the bad old days, with a pen and paper, for each final page of a manuscript I would write several pages of rough drafts. Now, with a word processor, rewriting is incredibly easy. No longer do I agonize over moving words, sentences, or paragraphs around; with the "block move" function, I do it with ease. Misspelled words do not bother me at all—they will be caught by my spell-check program.

Let us take a look at a typical word processor and see how all of these things tie together.

THE NUTS AND BOLTS OF WORD PROCESSING

The first step to word processing is to load the word processing program into the computer. The user can then either create a new file or review a file that has already been created. With some word processing programs, the program puts the user right into the page—it is like sitting down with a pen and paper. WordStar loads with its main menu and list of files ready for your selection.

The next step to word processing is to enter the document. Entering the document is just like using a typewriter. There are some minor differences, however, from using a typewriter. First, the word processor is slightly faster than a typewriter. The mechanical drag of a key striking a paper actually slows down the input, even with the best of typewriters. Second, if you make an error, there is no need to reach for the correction fluid; with a word processor simply press the delete key to erase the incorrect letter, or strike over with the correct letter(s).

After the file has been entered (or "keyboarded"), it is always a good idea to check spelling. Most word processors contain a dictionary which allows the user to check for spelling errors. Words can be added to this dictionary to meet the needs and vocabulary of the particular user.

That is all there is to basic writing. As mentioned earlier, writing is rewriting. The key to good written communication is heavy editing—and that is the true glory of word processing. Editing is simple. While reading over a first draft, if you spot a misspelling, simply use the arrow keys to move the cursor to the mistake. The correction can be made by either overwriting, by inserting or by deleting. If the whole word is wrong, you simply delete it and put in the correct word. If an entire line is incorrect, it can also be easily removed with one keystroke.

The best advantage of a word processor is full text editing. Whole paragraphs can be moved or deleted with only a few keystrokes. If there is a thought or a paragraph which needs to be moved to a different location within the document, this can be done with only a minimum of keystrokes. Some word processing systems have the capability to move sentences or paragraphs to a completely different document. Some also have a merge feature which allows the user to insert variables into a standard letter or document without having to keyboard the entire document each time.

All this is easy as long as the program makes it easy. These are very complicated functions you are asking your system to perform. A very sophisticated program can make these operations easy and simple.

The program must interface with the user in a way which makes the program almost invisible — it is the work that is important, not the means used to get it done. The ideal word processor is never obtrusive, never presenting problems, and always presenting solutions.

The top-of-the-line word processor programs, for home use, cost around $500. Less expensive systems are available, but should be studied carefully to ensure that they meet the demands of the particular user. The nature of our work calls for a quality system. A word processor without all the needed features will be a constant aggravation.

Resident Programs

Resident programs are available to change any word processing package to the unique needs of the user. A resident program is a program that can be operated while you are performing other functions on the word processing system. One or more resident programs are needed in almost every word processing system currently on the market.

A pop-up calculator is a resident program that is indigenous to the WordStar word processing system. The pop-up calculator can be accessed by a particular combination of keystrokes. Addition, subtraction, multiplication and division can be performed without leaving the document you are currently working on, and the answer can be entered directly onto the page. If a great deal of statistical work is done on the word processor, this resident program would be a tremendous asset.

Another resident program that could be helpful is an alarm. With this program an alarm can be set to go off at designated times throughout the day. This function would benefit a person who has a great deal of appointments during the course of a day.

Another useful resident is a printer controller which changes the type style of a dot matrix printer. The dot matrix printer is relatively inexpensive, can print at varying speeds, and has a wide variety of type styles available. Usually, these styles are controlled by switches on the side of the machine. With the resident printer program any print style can be set up without the use of the mechanical switches.

Another resident program that is available is the Pop-notebook. The Pop-notebook allows the user to make notes unrelated to the particular document that is currently being worked on without leaving that document.

A print spooler or buffer is another resident program that is very useful. A print spooler allows the user to send more than one document to the printer at the same time. The buffer accepts the file much faster than the printer can. The print buffer saves the file in memory until the printer is ready to receive it. This allows the user to use the word processing system while the printer is being used. Without the print spooler or buffer the user would have to wait until the document was printed before continuing with other work.

These are just a few of the resident programs that are available. There are many other resident programs that can help meet the needs of a particular user.

Nonresident Programs

A nonresident program is one that cannot be used while other work is being done with the system. The nonresident programs that I have found helpful include:

> *PC-Read*: This is used to measure the reading level of the material that has been written.

> *Check-disk*: This tells the user how much space is left on the disk that is being used to save documents.

> *Grammar and Punctuation:* Currently, grammar and punctuation evaluation programs are available only for CP/M machines, but several companies are planning to release new versions for IBM compatible computers. Grammar and punctuation programs scan your document one word at a time. These programs look for redundancies, vague wording, trite phrases, and improper sentence structures. Essentially, these programs check to see if you write according to a certain style. The user does not have to use the suggestions the program offers.

RECOMMENDATIONS

New word processing systems and options are constantly being developed; however, a powerful or more expensive program is not necessarily the best one as it may be excessively complicated and difficult to learn. In order for a word processing system to be effective it must meet the needs of the user.

A word processing system should contain the following features:

- *Good Documentation:* Word processors are complicated. A well written manual will make the learning process easier. It will also give you an idea of the program's interface—if the software company cannot provide a clear, well prepared manual, then it is unlikely that the program will be easy and simple.
- *A Choice Between Command and Menu Drive:* The menus will give the user the information needed to do a specific operation, but once these commands have been memorized the user will not want to go through the menus each time. A combination of both is best, with the menus appearing only after a few moments' hesitation, which tells the program that you need the menu.
- *Footnote Capability:* The user should be able to choose between having the same footnote on each page, or a specific footnote inserted on a particular page.
- *Ease in Operation:* The system should be easy. The ability to move sections around and renumber sections and pages automatically should be a part of the word processing program.
- *Search and Replace:* Search and replace is the ability to replace a specific word or phrase with another word. This is very useful if, for example, you have decided to change methyl chloroform to 1,1,1-trichloroethane.
- *File Length:* There should be no limit on the length of the file (up to the space available on the disk). Some programs will not handle files of more than ten pages.
- *Printer Controls:* The program should include printer controls. These include the ability to print superscript and subscript characters as well as the ability to print in boldface or to print underlines.

While looking for a word processor, plan to spend some time at the sales room using a demonstration copy. Use as many of the functions as possible. Find out if you can take it home for a few days (some retailers will loan or rent copies of the program to qualified potential buyers). Only by working with the system will the user know if it is right for his or her needs.

Another approach is to obtain a copy of a shareware word processor (such as PC-WRITE, for IBM-PCs and compatibles) and work with it for a few weeks. The shareware program can be obtained for a nominal copy fee from users' groups, or downloaded from CompuServe's IBM forum. While these programs do not fill the needs of industrial hygiene report preparation, they are available for a nominal cost (copy fees or download time, plus a small fee mailed to the author) and will help educate the new user about word processing. After using the shareware program, the user will have a good understanding of his or her specific requirements. The user will be in a better position to evaluate commercial word processors.

Learning to use a word processor is an investment of time and money. The investment is repaid with substantial dividends. Report preparation becomes much less time consuming, which increases productivity in other areas. The quality of communications also improves. In a profession such as industrial hygiene, where complex, technical concepts must be conveyed to readers who generally have a limited knowledge of what we do, anything that improves communication is worthwhile. The payoff, then, is indeed worth the price.

CHAPTER 11

COMPUTER COMMUNICATIONS

CHARLES M. BALDECK, PhD, CIH
Industrial Hygiene Specialties Company, Columbus, Ohio

SOME DEFINITIONS AND PURPOSES FOR COMPUTER COMMUNICATIONS

Computer communications can be thought of as simply the transfer of information from one computer to another. Information is stored on computers in "files" of either 8-bit (binary) or 7-bit (normal ASCII) characters. It is transferred by sending these files, one character at a time, from one computer to the other. The transmission speed is called "baud rate" (1 baud = 1 bit/sec), and the overall process is called "uploading" when the transmission is from a computer (usually a micro) to the "host" (often a mainframe), and called "downloading" if the information flow is from the host to another computer.

Conceptually, think of potential computer communication applications falling more or less into one of two camps: information storage and retrieval (data base) operations or networking. An example of the former would be using the microcomputer to extract desired information from a larger mass of data, such as the National Library of Medicine files, which is stored in some logically organized fashion at a physically remote site. This kind of process typically involves one human operator using his/her microcomputer and a communications link to operate as a remote terminal of the larger computer where the desired information is stored in a formally organized fashion. Networking, on the other hand, can be thought of as using the microcomputer to mediate an exchange of information between two or more people, who are using it as a tool of personal communication in much the same way that they might exchange conversation, written letters or telephone calls. In other words, networking deals more with information stored in the minds and private files of the participants, than in formally organized data bases. Both information retrieval and networking applications can be powerful tools to enhance the productivity of people like industrial hygienists, who work in settings where timely access to information is an important factor.

Using computer communications for remote data retrieval is probably the most immediately appealing application for most users. There are several important data bases available already containing the kinds of information which are useful professionally. For example, microcomputers can be used to call up and search for specific information from the collection of Chemline, Toxline, Toxicology Data Base, and RTECS files maintained by the National Library of Medicine, or the multiple data bases maintained on the Lockheed Dialog system. Another similar application might involve retrieving MSDS information on specific chemicals, either from the manufacturer, or from commercial information services such as Chemical Abstracts or Hazardline. The Mallinckrodt Company now keeps MSDS information for their products on a modem-accessible microcomputer, available 24 hours a day. This service proved to be so popular that Mallinckrodt had to expand their system to accommodate all the inquiries. Word has it that J.T. Baker, DuPont, and Fisher Scientific will be making similar access facilities available shortly for their products.

Many users would like to be able to access some of these information sources in an efficient and cost-effective manner, but are uncertain of how to get started. Computer communications has always been one of the harder applications to get "up and running" smoothly. This chapter will try to take some of the mystery out of it, by reviewing the basic equipment, concepts, procedures, terminology and other general considerations which are common to all types of computer communications. Although this chapter does not go into specific detail on individual systems the user may want to "call up," it will attempt to point out some common features where appropriate. Finally, this chapter will examine where computer-based communications appears to be headed, and some applications which can be of significant future value and interest to industrial hygienists.

THE BASIC EQUIPMENT

The basic elements required to go "on-line" for an information transfer are (1) a "host" computer, (2) a microcomputer, (3) some type of communications software to control the transfer, and (4) the connection between the two computers.

Generally, it is felt that anything with less than a 40-column display will not be very practical for most work. However, almost any microcomputer can be used for communications with another machine. Light weight (4–5 lb) "lap-top" portables with a 40-column display, such as the Radio Shack Model 100 or the NEC PC8201, can be used where small size, light weight and portability are primary considerations. Where weight and size are not as critical, a "minimum" machine with an 80-column display, 2 floppy disk drives and at least 64K of memory would be recommended. Almost any of the larger portables and desk-top machines intended for serious business use would qualify.

SOFTWARE

There are many communications programs available, the selection depending on the microcomputer being used. Some of the more popular ones are ASCII Express, Apple ACCESS and Hayes Smartcom I, designed for the Apple II; Red Ryder for the Macintosh; and for CP/M machines, Crosstalk, MEX (PD) and Modem-7 (PD).

For the IBM-PC and compatibles using MS DOS, one can select from many different levels of power and convenience. The most powerful are state-of-the-art programs which include a script language, allowing them to be configured for highly automated and even unattended operation. Programs in this class would include Microsoft ACCESS, Crosstalk Mark IV, Relay Gold, ProYam and PIBTERM (a "shareware" program). Other very capable programs, with macro capability but without a complete command language would include ASCII Pro, Crosstalk XVI, MEX-PC ($59), and ASCOM IV. Some "integrated" programs, such as Framework, Enable, or Symphony, also have communications "modules." Although these are seldom as powerful as the stand alone programs, they may be perfectly adequate if the user prefers the "integrated" software approach for other applications. For the IBM-PC and MS DOS compatibles, there is also a large and growing body of "shareware" communications programs and utilities, which one might wish to explore. The shareware concept encourages the user to copy the program, try it out, and then leaves it up to the user and his/her conscience to send the author a nominal requested contribution if the program is found useful. Shareware is a good way to try several programs at low cost before adopting one for permanent use. Some of the better known shareware communications programs are PC-TALK III (and its many derivatives and modifications), Q-Modem 3.0 and PIBTERM 3.2.5. Later versions of these may be current by the time this chapter is read.

THE CONNECTION

The connection between a micro and another computer can be either a direct cable link between the serial interface ports on the two machines or a modem transmission. If the two machines are in close physical proximity, the direct wire connection is relatively simple, works well at speeds of up to 9600 baud and for runs up to 50–100 feet or so. A shielded cable should be used for best results to prevent electromagnetic noise interference from garbling the transmission. This is a sample and quick way to transfer files from one microcomputer to another when the disk formats are otherwise incompatible. If the host machine is in a different building or city from where the micro is located, a modem connection is used. A modem is a device which translates the electrical "bit" pulses into audio frequency tones, which can be transmitted over telephone wires. There must be two modems, one on each end, for a modem connection, one to send the tones and one to translate these tones back into electrical signals for the receiving computer. They also must be able to "talk" to each other.

There are many different types of modem and transmission line combinations which can be used. High-speed modems (and modem multiplexers) are available which can operate at 9600 baud, but reliable data transmission at these speeds requires specially conditioned and balanced data grade telephone lines, which must be leased from the telephone company. These are more expensive than ordinary voice lines, and are used where large amounts of data must be routinely transmitted between fixed locations.

Most microcomputer communications are done through the switched public telephone network (ordinary voice grade lines). While considerably cheaper to use, there are often "noise" problems associated with voice line connections, which may became worse as distance from a substation increases. Voice circuits were not designed for data transmission and connections can be quite poor (glitches), and still be acceptable for voice transmission. The telephone company states that it "will not guarantee the quality of data transmission over voice lines." In practical terms, this means if the user has transmission problems with a noisy line, the phone company will not respond on that basis, except to suggest leasing a data line from them. To get any service on the noise problem the user will have to disconnect all of the computer equipment, reconnect the telephone set, and tell customer service that the telephone line is too noisy.

A third alternative, available for connection with some hosts, is the use of a commercial data packet switching network such as Telenet, Tymnet, Uninet, or Datapac (in Canada). Connection is made through a local access number, which routes the call through its own network to the host. Identification of the terminal type and the access code for the host computer being called is required. The packet switching network invoices the host for this service, and the host in turn includes these fees in its access charges to the user. Therefore, this option can be used only if there is a contractual agreement between the host called and the packet switching network; this option cannot be used to call another private microcomputer in a distant city.

MODEMS AND HOW THEY ARE CONNECTED

A modem can be mounted inside the computer (internal) or contained in a separate "box" beside the computer (external). Both types perform the same function—connecting the computer to the telephone line. An internal modem is more compact and out of the way; there are fewer external cords and connections involved. An internal modem may be best for portable computers, where everything is compact and in one unit. Internal modems are constructed on a circuit card which plugs directly into a "buss" connector inside the computer. They are generally available only for computers with such a buss, e.g., the Apple II+, Apple IIe, or IBM-PC and compatibles. Also, an internal modem can only be used with the specific computer model for which it was designed. Consequently, if the computer is sold later, the modem will probably have to be sold with it.

An external modem (sometimes also called an RS-232 modem) can be connected to any computer through the RS-232 serial port which nearly all computers have. An external modem can therefore be moved readily from one computer to another and there is a wider selection of models and features. Since the modem does not have to physically fit inside the computer "box," there is more room for the designers to add hardware support for extra features.

Although the RS-232 serial port is supposed to be a "standard," it is really not standardized enough to be sure that if any RS-232 device is plugged into any RS-232 port, it will work. To connect a modem to a RS-232 port, two facts need to be known about the computer's serial port whether it is a "DTE" (Data Terminal Equipment) or a "DCE" (Data Communications Equipment) port, and its gender (male or female). If the computer port is configured as a DTE port (most are), a "straight through" cable can be used, with appropriate gendered plugs on each end, and pins 1–8 and 20 connected (pins 12 and 22 may be optionally used for some modems). If the serial port is configured as DCE, then a special cable is needed to connect the port to the modem.

Smart and Dumb Modems

Modems are sometimes referred to as "smart" or "dumb" modems, and as either "direct connect" or "acoustically coupled." The D.C. Hayes Company, which made the first popular ASCII-controlled modems, coined the term "Smartmodem" to distinguish the fact that they were controlled by software commands issued as ASCII character strings sent by the microcomputer. The term is now used to refer to any ASCII-controlled modem. Internal modems are generally "smart" modems. The corresponding term of "dumb" modem refers to modems which are not ASCII-controlled.

The term "direct connect" refers to a modem designed to be attached directly to the telephone line in the same way a modular telephone is connected; by plugging it directly into a telephone extension cord. Acoustically coupled modems, on the other hand, have a pair of rubber cups housing a small microphone and speaker, and the telephone handset is placed over these cups to establish an acoustical connection through the telephone set (much like calibrating a noise meter). The most popular modems made now are of the "smart" and direct connection variety.

Another term describing modems is "Hayes compatible." This refers to the standard set of 20 commands and 6 responses which were incorporated in the original Hayes Smartmodem 1200 modem. The responses are switch setable to return results either in English or as numbers, e.g., "CONNECT" or "1". The Hayes Smartmodem was so popular that its command set has became an industry standard, and most present day software expects to work with the standard Hayes command set. Some modems have a superset of the Hayes commands, but can be set to operate exactly like the Hayes, if this is required by the software. These are acceptable, and may give more features for the money, but do not purchase a "smart" modem which is not Hayes compatible. The software may not work properly unless the user "patches" it, which is not an operation for beginners.

There is another, different type of "smart" modem, which appeared in late 1985. This deserves mention because it may represent the direction of future modem developments. These special modems feature very sophisticated circuitry to achieve both dynamic error checking and correction and very high transmission speeds over voice grade lines. An example is the Telebit Corporation's "Trailblazer" modem. The cost is high: $2000 (internal) or $2400 (standalone). It uses a 68000 microprocessor chip and dynamically monitors the condition of the telephone connection, using all of the available bandwidth to transmit and receive at speeds up to 8000 baud, error free, with incremental fallback in case of line quality deterioration. It only works in this mode with another modem of the same type on the other end of the line. It also has a superset of the Hayes commands and can work at 300 and 1200 baud with ordinary modems (but without the special error correction or automatic incremental fallback features). The high speed mode is not supported presently on any bulletin boards or public network services and requires software which can support the speed (Crosstalk-Fast Ver. 3.51, PC-TALK III-B, Pro-Yam for the IBM-PC; Macterminal and MITE for Macintosh). A somewhat exotic product, however, even at current prices it could be cost effective in certain applications. If the price comes down and the technology finds widespread support, it could become a significant factor in the future of modem communications.

The "dumb" or non-ASCII controlled modems have fallen out of favor with most microcomputer users, although there are still quite a few of them around. Some are direct connect, some are acoustically coupled through a telephone receiver, and a few can work either way. They are much less complex to understand and operate than "smart" modems, and may actually be better suited to certain applications, such as use with a wide variety of equipment, as when travelling. To use a "dumb" modem, the user must manually dial the number of the computer, flip a switch from "voice" to "data" when the high-pitched squeal of the answering modem is heard. When finished, the switch must be thrown again to hang up. Some will auto-answer and some will not. Some have a manual switch to select originate or answer frequencies. "Dumb" modems can be obtained sometimes at a very good price, because the market has moved mostly to the ASCII-controlled modems. For those on a tight budget, a good "dumb" modem may be a better buy (cheaper and more useful) than a non-Hayes compatible "smart" modem. Nearly all communications programs offer some support for a "dumb" modem.

Modem Frequency and Modulation Standards

For modems to communicate with each other, they must use the same "standard" set of frequencies and modulation protocols. In the United States these standards (sometimes called transmission protocols) have been dominated historically by the presence of AT&T on the communications scene, and have been mostly the standards established by Western Electric for the modems which they build for AT&T.

The important "Bell" standards in nearly universal use now in the U.S. for two-way asynchronous communications are:

1. Bell 103: up to 300 baud, full or half duplex.
2. Bell 212A: 300 or 1200 baud, full or half duplex.

For ordinary use on the dial-up telephone system in the U.S. and Canada, most of the modems will be operating on one of the Bell standards—Bell 103 for 300 baud or Bell 212A for 1200 baud. The Bell 212A standard is also fully compatible with the Bell 103 standard, and is therefore the most popular type. Do not confuse the latter with the Bell 212 (without the A) which operates only in half duplex at 1200 baud. There is also a Bell 202, which is 1200 baud half duplex only, but this is not commonly used.

There also have been some non-Bell standards, one of which was the Vadic 3400, for 1200 baud. However, when Bell came out with the 212A standard, the Vadic became obsolete, although there are some mainframe systems which still support Vadic as well as Bell protocols.

Outside the U.S. and Canada, modems use frequency and modulation schemes established by treaty through the International Telephone and Telegraph Consultative Committee. The CCITT standards are different than Bell standards for 300 and 1200 baud; therefore, European modems theoretically cannot "talk" to U.S. modems. Even so, U.S. 1200 baud modems use frequencies which are close enough to European standards that they will sometimes work. For 2400 baud operation, the CCITT standard was adopted in the U.S., making it universal; consequently, 2400 communication should be compatible everywhere.

Telephone standards and regulations outside the U.S and Canada vary a great deal. Many countries have severe restrictions, taxes, fees, licensing laws and other "red tape" on the use of modems. If computer communications from outside the U.S. is desired, the user should check out the situation ahead of time.

Modem Settings

Most modems have a number of configuration options which must be set, either with physical switches on the modem or by software commands sent from the computer. For the latter type, the switch settings establish default values, and the user has the option of overriding them at any time through software commands. These are explained in the instructions which come with the modem. A "dumb" modem may only have a Voice/Data switch and an Originate/Answer switch. The operation of the Voice/Data switch was discussed above. If the modem has an Originate/Answer switch, remember to set it on originate when calling, or the answer mode when receiving a call from another computer.

GETTING ON-LINE

Most of the problems encountered with computer communications involve the initial setup and operation. Once this is done, future operation is quite easy. If problems develop during initial setup, ask for help from someone who has already gone through the process, or join a local computer club and ask for help. Mutual support among the microcomputer user community is usually quite good.

First, installation or configuration of communications software for the modem and computer is necessary. This tells the computer about the equipment setup and sets the program defaults. If in this process the program asks a question the user does not understand, it should be written down, and the user should select the default answer (or guess) until help can be obtained from the dealer or the program/modem support group, or from someone at a local computer club. Make notes of all the modem settings and software options selected. As more is learned about modems and software, the user may want to change some of these settings for greater convenience, or to establish the ones used most often as the boot up defaults.

For each host called, the user will need to find out and set the communications program for baud rate, full or half duplex, the "data word format" and any required terminal emulation. The baud rate will depend on the capabilities of the modem and the host being called. It is nearly always 300, 1200 or 2400 baud. Some host systems charge a higher rate for connection at the higher speeds. If the user has a choice, the higher speed should be selected for uploading or downloading files, and the lower speed for reading text on the screen or typing. Using full or half duplex depends on the host. Full duplex is used for most public networks and bulletin boards. When typing, if nothing appears on the screen or the letters are doubled LLIIKKEE TTHHIISS, try the opposite duplex setting. The "data word format" refers to the number of data bits, stop bits, and parity. The most common is 8 data bits, no parity, 1 stop bit (8N1), followed by 7 data bits, even parity, 1 stop bit (7E1). One of these will cover about 99% of the cases encountered. There are other data word formats, but they are quite seldom used.

If connected to a mainframe which expects to use a certain type of terminal, the user may need to use a terminal emulation mode. These modes let the computer imitate the expected terminal (VT-100, VT-52, IBM 3101, etc.). Terminal emulation is not generally required for most public networks or bulletin boards, although it can often be used. If asked for a choice of terminals to emulate, select the "ANSI," "dumb terminal" or "CRT" settings for these.

Once the software has been configured, dial the telephone number of the computer to "talk to." If the user has a "smart" modem, it will dial, listen for the answer tone and connect automatically. Once connected, the user usually must send a recognition character (usually 1–3 C/R or a control-c) to "wake up" the host. The remote system will then prompt for some combination of name, I.D. or password to identify the author as an authorized user. This brings up a very important issue.

PROTECTING THE PASSWORD

Everyone has heard about "hackers" who use modems to break into computer systems and cause havoc. This has increased awareness about the need for computer security. There should be a separate password for each of the different systems with which you communicate.

CAUTION: TAKE PRECAUTIONS TO SECURE YOUR PASSWORD FROM THEFT AND UNAUTHORIZED USE.

When typing in the password, tell others to look away, so they cannot see what is being typed. DO NOT use a name, initials, a child's or dog's name, or some other easily guessed word for a password. The most secure passwords consist of two or more unrelated words joined by a non-alphanumeric character such as * or %, e.g., BOAT%TOUCH. NEVER give the password to someone else, and change it periodically, just in case. If you have saved disks with auto log-on routines in the communications program, the password will be on those disks. Take steps to secure them, and do not give out copies of working disks. If copies must be given out, give the unconfigured (distribution) disks only.

Be alert for any "con" attempts. These attempts often take the guise of simulated "system problems." If someone may be doing this, report it at once to the system operator. Such incidents are rare (because system operators kick those who try it off their systems) but they have happened. When logging onto a remote system, the password is the one item which should NEVER show on the screen when it is typed in. If it does, something is wrong—change the password immediately and report the incident. Be alert for any unusual messages or prompts during the sign-on routine. If anything is suspicious, hang up and try again, or call customer service.

Enter the password ONLY at the initial system sign-on routine, or when changing it under the NORMAL password change routine provided by the host system. NEVER REENTER THE PASSWORD AFTER BEGINNING TO USE THE SYSTEM, especially if it is a large system with many other users. If the system appears to be having problems and requests the password after initially logging on, DO NOT GIVE IT. Instead, report the incident to the system operator immediately.

Remember, the user is responsible for all charges accumulated by the password used. It is like a blank, signed check. Do not let someone get it and have the opportunity to charge several hundreds (or thousands) of dollars to your account.

DEALING WITH TRANSMISSION ERRORS

Any data transmitted over phone lines can be corrupted by line noise and various glitches peculiar to long distance transmission technologies. Since even one displaced bit in a program file can make it useless, or one missing or extra character in a critical file (such as a company's payroll) can cause major problems, various methods have been developed for checking the integrity of data transmission.

Data checking and correction methods, called "protocols" operate either at the "link level" or the "file level." A link level protocol works full-time on the communications link, checking every bit of data transmitted, whether it happens to be in a critical file or not. File level protocols are designed to check just the contents of critical files which are transmitted. Although the operation may be more or less transparent to the user, file level protocols are invoked typically just before transmission of a critical file is started, and are terminated when the transmission of that file is finished.

The most popular and widespread file level error checking and correction protocol is the X-modem protocol. It was developed by Ward Christensen, who generously placed it in the public domain. It is very widely supported by most micro- and many minicomputer systems, but not by most mainframe systems. Under X-modem protocol, files are divided into successive 128-character "blocks" which are then transmitted along with the corresponding block number and checksum or CRC results. The receiving computer also does the same calculation on the received block, and checks to see if the results agree. If they do, then it accepts that block, if not it requests that the block be resent. If there are too many errors in a row, the transfer is aborted. There are many variations and slightly different implementations of X-modem protocol. The original error detection algorithm was a checksum, which will catch about 99% of the possible errors. Many current implementations of X-modem now support CRC error checking. This is a more sophisticated and reliable method.

Modem-7 is a variation of X-modem which provides for sending a batch of files at one time, each one preceded by a file name. It can use either checksum or CRC error checking protocols. Y-modem is another variation which uses 1024 character blocks. It is more efficient than X-modem for transfer at speeds of 2400 baud and higher. There are single and batch file versions. Telink is another variant of modem-7 that adds information on file size and creation date to the file name. Telink was written by Tom Jennings, who wrote the software for the Fido bulletin board system, therefore Telink is most useful on Fido systems.

Kermit is another popular file level protocol. It was developed at Columbia University and is in the public domain. Kermit is widely supported on micros and probably the most common and widely available protocol on mainframes. It is used in micro to mainframe links in universities and government, and can transmit one or a group of files. There are three different error checking protocols used in Kermit implementations: (1) 6-bit checksum, which is allowed by all Kermits, (2) 12-bit checksum, which is supported by some Kermits as an option, and (3) 16-bit CRC, which is supported on some Kermits. Choose the latter one if the host will support it.

There are also some proprietary file level protocols, which were designed only for use on a particular host system. Examples are the CompuServe "A" and "B" protocols, which are supported by some communications programs. Although such protocols are usually very convenient to use, their usefulness will depend on how much the user intends to use the hosts for which they were designed.

The "link level" protocols are in less common use, because most of them are proprietary. They can be used with some packet switching networks. Two of the more common ones are the "X.pc" and "MNP" protocols. X.pc is supported by Tymnet and sends all data in "packets" which are individually routed to the proper host and checked for accuracy. It has the unique feature that it can support up to 16 concurrent sessions between a micro and different hosts. The MNP protocol is proprietary to Microcom, although it is licensed to others. It is supported on Telenet and Uninet, but does not allow concurrent sessions.

For noncritical ASCII files, one can often do a satisfactory text transfer with no protocol in effect at all, but a noise-free connection (like the serial cable wire connection) is needed for best results. This process is often called "buffer dump," "DC2/DC4," or "straight ASCII file transfer," and is often listed on menus as a protocol, although it is not really a protocol in the sense that any error checking is done. This process typically uses X-on (control-Q) and X-off (control-S) "handshaking" for flow controls (to avoid overrunning the buffer on the receiving computer) but has no error detection or correction. Depending on the quality of the connection, occasional characters are dropped, or extra "garbage" characters are transmitted from line noise. Do not try to use this method for the transfer of 8-bit files, programs, or critical data. For those a real protocol is needed.

Although nearly all computers recognize the X-on/off control codes, some word processors do not. When using the "straight ASCII file transfer" to download files from a word processor or other device, one which does not support flow control, be sure not to send more text in one continuous block than the communications buffer can hold. If the communications buffer has to write out to the disk, characters will be lost.

SOME TIPS FOR EFFICIENT USE OF ON-LINE TIME

Many host systems charge according to the time spent connected to them. The steeper the time charges are, the more it pays to minimize the "connect time." Develop operating procedures which will log the user on, do the work to be done as quickly and efficiently as possible, and log the user back off. Here are a few tips to help get "up to speed" quickly on a new system.

For the first few times on a new host, open the communications capture buffer before logging on and capture the entire session to disk, so it can be printed out for more detailed review later. It is possible to edit the file with a word processor to save and print out all the menus, bulletins, help files, and command summaries. Review these and have them handy for the next time.

Learn how to download and upload to the host system. If it is a "message" system, "Read" messages and files by downloading them onto a disk for later perusal. Then compose replies, E-Mail messages, etc., off-line using a favorite word processor. Save them to disk as ordinary ASCII text files without the control and special formatting characters (which some word processors insert into the text of saved word processing files which are destined for printing). Then upload the messages, replies, etc., from disk files using a communications program. Use straight ASCII file transfer for text. It is generally faster than a protocol transfer, and the latter is not necessary unless the information is critical or the line noise is excessive. If the line noise is too bad, producing weird characters on the screen, it will have the effect of slowing down a protocol transfer. In such cases, it is usually best to just hang up and try again later.

After becoming familiar with the operations on a frequently used host, try automating all or part of the frequently used operational sequences by using "macros" or "scripts" to save keystrokes and time while on line. There are even special programs to automate entire sessions. Macros are used to assign a sequence of frequently used keystrokes to a single key, to save the user repetitive typing. Many communications programs already support macros, but if it does not, they can be obtained by using a resident macro key program (like Borland's SuperKey) to define them. Scripts are more powerful than macros and are really miniprograms within the main communications program to automate a whole sequence of different operations. Only some communications programs support scripts, but there are at least two good shareware/PD programs which do: Q-Modem 2.0 and PIBTERM 3.2.5.

Separate programs also can be written in BASIC or another language to automate interactions with remote systems. There is even a special language, called MIST+ (New Era Technologies, for IBM-PC and compatibles, cost around $500) which is designed especially for that purpose. Extremely sophisticated computer communications applications can be written in MIST+.

For CompuServe users with an IBM-PC or compatible, the use of electronic mail and messaging on the CompuServe Forums can be highly automated, using a very fine (free) program called AUTOSIG. AUTOSIG can be downloaded from the IBM-PC New Users Forum on CompuServe (Go IBMNEW) and it works splendidly on the SafetyNet Forum. The latest version is 4.10L, which is the one to get if using AUTOSIG for the first time. There is also a special version 4.12 which can be run concurrently under DESQview. With version 4.12 and DESQview the user can retrieve or send messages which have been prepared ahead of time in background mode while the computer is busy with another task. With an IBM-PC or compatible, get AUTOSIG and try it out, if only to get an idea of what can be done in automated computer communications. AUTOSIG was written in BASIC and then compiled. However, with a communications program with a powerful script language, a script program can be written fairly easily to do the same kind of thing for nearly any host system.

HOW WILL COMPUTER COMMUNICATIONS AFFECT INDUSTRIAL HYGIENE PROFESSIONALS?

Computer communications applications can be divided into information storage and retrieval (data base type) operations and networking applications. Networking applications, such as electronic mail and bulletin boards, are probably less familiar to most industrial hygienists than the data base applications, such as searching Medline or Dialog. Although the use of computer communications to access remote data bases will continue to be important, networking, presently in a very early stage of application in the industrial hygiene profession, ultimately may have an even greater impact. This development will be stimulated by (1) the better communications "tools" which will be available, and (2) a generally increasing urgency to improve working efficiency.

In the most general terms, the effective practice of industrial hygiene depends very heavily on the timely acquisition and dissemination of knowledge and information, much of which changes or needs to be updated frequently. Basically, industrial hygienists are specialized "knowledge workers" (i.e., work more with heads than with hands), and it is difficult to see how that fundamental situation will change in the foreseeable future. This means that adequate information flow is, and will continue to be, absolutely critical to professional success. It follows then, that any systems which make the information flow more efficient will sooner or later recommend themselves to our attention. They will do so because of the pressure to be more productive.

As computer communications technology matures and gets easier and more "time-efficient" to use, it will inevitably win wider acceptance, not only with industrial hygienists, but also among the broader business community which industrial hygienists deal with daily. There should be developments which improve the communications "tools": (1) better standardization, both in hardware and in human-software interfacing; (2) more widespread support and use of link level protocols, so that error-free transmissions can be pretty much taken for granted; (3) higher transmission rates with a matching sophistication of error correction (perhaps as exemplified in the "Trailblazer" modem system); and (4) standards and support for the transmission and receiving of images (graphs and pictures).

ELECTRONIC MAIL AND CONFERENCING

As these improvements become available, we will see a much greater use of the computer communications "networking" applications which are aimed toward facilitating the exchange of information between professionals, greater use of electronic mail for one-to-one messaging, and electronic bulletin boards for all kinds of one-to-many and many-to-many business and professional communications. This will be facilitated by automated sending and receiving (as in AUTOSIG) with many of these applications running as background tasks in a multitasking environment such as DESQview. The computer will be able to send and receive these messages while the user is doing something else.

Most electronic bulletin boards combine an electronic mail function for private communications with a public messaging capability. The public messaging can be used to hold "conferences" in which participants sign on and read previous transactions, then respond with their own input and questions, checking back in later for the responses. When done in this manner, it is called "extended conferencing" as the conference extends over several days or weeks, and each person participates intermittently at their own convenience. Some systems, however, also have the additional ability of "real-time" conferencing, in which all of the participants are on-line at the same time. This mode is much like a physical meeting and someone, analogous to the chairman of a physical conference, has to settle questions of precedence and keep order. Depending on the system, there can be several such groups working in different conference areas at the same time, and because of the way the system operates, one individual can "monitor" a number of simultaneous sessions, contributing comments to each of them in turn. Finally, both real-time and extended conferencing can be combined, according to the business at hand.

Because computer-based conferencing can be much more efficient in terms of time spent versus business accomplished, it may eventually replace many conferences presently carried out by physical meetings and phone calls. Computer conferencing can save travel time and cost by eliminating the travel. Why waste several hours traveling (not to mention the expense) just to spend a few hours doing business, when nearly all of that business could be done through computer-mediated conferencing? One very big advantage for busy people is that messages can be sent and received at a time convenient to each participant, minimizing the problems of conflicting schedules. For a complete review of conferencing systems and applications, see the December 1985 issue of *Byte* magazine.

Communications by either computer-mediated conferencing or messaging can be more efficient than voice telephone calls for some types of communications. It eliminates the time and aggravation wasted on unproductive "telephone tag." The deliberate act of writing out a message encourages more thought about exactly what is being "said" and how, thereby enhancing the potential value of the message, and a written record of all communications and transactions is always available for review. Some busy executives in other professions, who are accustomed to the advantage of computer-mediated communications and who wish to conserve their available time and energy, will volunteer for or chair a professional committee only on the condition that its business is conducted electronically. We expect that such thinking will eventually percolate into the professional industrial hygiene community as well.

The idea of a national electronic bulletin board dedicated to industrial hygiene issues has long been one of my own special interests. Although such a system has the potential to be the full information equivalent of a national conference, it has the additional advantage of being available to any and all of the participants at any time which is convenient for them, and at a fraction of the cost, without their even having to leave home to participate. Some day these capabilities may be exploited to create the electronic equivalent of a national professional conference.

The only attempt to actually create such a network for the exchange of industrial hygiene information was the On-Line Industrial Hygiene Forum, which began in early 1985 on the CompuServe information service. Although it was started as a strictly industrial hygiene-oriented network, disappointing acceptance and participation by the industrial hygiene community during its first year of operation finally led to merging with another group, broadening the subject focus and renaming it as the "SafetyNet" Forum. On SafetyNet, all kinds of safety-related issues besides industrial hygiene, such as hazardous waste, preparation for chemical emergencies, nuclear safety, fire prevention, etc., are discussed.

Although the On-Line Industrial Hygiene Forum was apparently a few years too early for most of its potential audience, the idea of a professional network has considerable potential to "catch on" as more people inevitably discover the practical, social, and professional values of electronic networking. Practically, a professional network is a way to identify and communicate with others having similar special interests, get help with specific problems, and exchange specialized information among fellow professionals. Socially, it is an ideal medium for informal contacts, offering mutual support and friendship among peers. Many strong friendships are formed among regular participants on a message board. Professionally, it can be an effective enhancer of professional recognition and reputation; by sharing knowledge generously and freely with others, people quickly learn who others are and what they know. For example, reporters often use computer-based professional networks to identify persons who can provide technical expertise for articles and stories on a specific topic. As a direct result of knowledgeable messages on the subject, left on a professional network board, one of our SafetyNet members was approached recently by a New York publisher with a retainer offer to advise them on chemical hazard issues.

COLLABORATIVE AUTHORSHIP

A networking application which is closely related to conferencing is the use of electronic communications as a tool of collaborative authorship. It is an ideal way for a group of individuals, such as a professional committee, to temporarily pool their specialized knowledge as co-authors and/or editors of special publications such as monographs, articles, and books on timely topics. Because all of the notes, outlines, draft chapters, revisions, criticisms, feedback, rewrites, and finally the completed work can be available to all of the participants at all times, with changes instantly available to all as soon as any one of them has added or revised something, the creative cycle of draft, review, feedback, and rewrite can proceed very rapidly, with everyone able to access the latest version of any part at any time.

The first industrial hygiene publication prepared by collaborative authorship over a computer communications network was done on the On-Line Industrial Hygiene Forum, and serves as a good example of what can be accomplished. This was an article about the Forum which was written interactively with its members by Leonard Wilcox, during the latter part of 1985. The speed of publication was remarkable: four successive drafts in 15 calendar days, including the Thanksgiving holiday. Comments and feedback from several commentators were incorporated into the text during that time, with the final version edited from the fourth draft and available for inspection by all commentators one day later. The final copy was simply left in the Data Library after completion, for the ACGIH editors to download when they were ready for it. The article was published in the premier (April 1986) issue of the new ACGIH journal, *Applied Industrial Hygiene*.

ELECTRONIC PUBLISHING

Electronic publishing is yet another possibility of computer-based communications. Strong candidates for on-line electronic publishing are: (1) summaries of not-yet-published current research, (2) new hazard information and alerts, (3) changes in government regulations, (4) comment solicitations for proposed regulations, and (5) other government (NIOSH, OSHA) information for which wide and rapid dissemination is an important consideration. The key word here is timeliness. There is no other publishing medium which can get updated written information to its audience nearly as quickly.

SECTION IV

Example Applications

CHAPTER 12

A LOTUS 1-2-3 DATA BASE MODEL FOR KEEPING AND TRACKING CITATIONS

BRYAN D. HARDIN, PhD
National Institute for Occupational Safety and Health, Cincinnati, Ohio

When reviewing literature for a publication on the reproductive toxicity of glycol ethers, it was necessary to organize information from a large number of publications dealing with many different glycol ethers, many test species, and multiple routes of exposure. To simplify organization and retrieval of that information, a Lotus data base was set up in which the key information and freeform comments could be entered (Figure 1). New references can be added as they accumulate to keep the data base up-to-date. Data fields are:

SEX: M or F — the sex of animal treated experimentally.
GLYCOL: The abbreviated glycol, glycol ether, or related chemical tested.
ROUTE: Route of ingestion.
DURATION: The days of gestation and/or total duration of experimental treatment.
MMOL/KG/DAY: The dose used for all but inhalation exposure. Conversion of doses to mass units (mmoles) allows direct comparison of the dose levels for different glycols. By making an entry for every dose level in multiple-dose studies, lines covering several studies of one glycol can be interleaved to develop a perspective on development of toxic responses over a wider dose range than is covered in a single report.
PPM: Airborne exposure concentrations in inhalation studies.
SPECIES: The test organism.

GLYCOL ETHER FEMALE REPRODUCTIVE TOXICITY

SEX	GLYCOL	ROUTE	GESTATION DAYS OR DURATION	MMOL/ KG/DAY	PPM	SPECIES	REFERENCE	OBSERVATIONS
F	EGME	ORAL	7 TO 14	18.40		MOUSE	67	100% INTRAUTERINE DEATH, 14% MATERNAL MORTALITY
F	EGME	ORAL	8 TO 15	0.41		MOUSE	47, 50	SKELETAL VARIATIONS
F	EGME	ORAL	8 TO 15	0.82		MOUSE	47, 50	TERATOGENIC
F	EGME	ORAL	8 TO 15	1.64		MOUSE	47, 50	TERATOGENIC, REDUCED FETAL BODY WEIGHT
F	EGME	ORAL	8 TO 15	3.29		MOUSE	47, 50	TERATOGENIC, INTRAUTERINE DEATH, REDUCED FETAL BODY WEIGHT AND MATERNAL BW GAIN
F	EGME	ORAL	8 TO 15	6.57		MOUSE	47, 50	98% INTRAUTERINE DEATH
F	EGME	ORAL	8 TO 15	13.14		MOUSE	47, 50	100% INTRAUTERINE DEATH, REDUCED MATERNAL WBC
F	EGME	ORAL	8...15 (1 DAY)	2.63		RAT	12	100% INTRAUTERINE DEATH
F	EGME	ORAL	8 TO 11	6.57		RAT	129	STAGE-SPECIFIC TERATOGENICITY, INTRAUTERINE DEATH, NO MATERNAL TOXICITY
F	EGME	ORAL	12 OR 13	3.29		RAT	129	TERATOGENIC, INTRAUTERINE DEATH, FETOTOXIC, NO MATERNAL TOXICITY
F	EGME	IP	12 OR 13	4.19		RAT	63	TERATOGENIC
F	EGME	IP	9, 11, 13, OR 15	2.49		RAT	4	TERATOGENIC, FETOTOXIC, INTRAUTERINE DEATH
F	EGME	DERMAL	7 TO 21	2.63		RAT	12	100% INTRAUTERINE DEATH
F	EGME	SUBCUT	7 TO 21	0.50		RAT	12	NO ADVERSE REPRODUCTIVE EFFECT
F	EGME	SUBCUT	7 TO 21	0.51		RAT	12	INTRAUTERINE DEATH, REDUCED POSTNATAL SURVIVAL
F	EGME		7 TO 21	3.17		RAT	12	100% INTRAUTERINE DEATH
F	EGMEA	ORAL	7 TO 14	10.37		MOUSE	59	100% INTRAUTERINE DEATH, NO MATERNAL MORTALITY
F	EGMEAcT	ORAL	7 TO 14	4.99		MOUSE	59	100% INTRAUTERINE DEATH, 30% MATERNAL MORTALITY
F	MAA	ORAL	7 TO 14	13.10		RAT	63	TERATOGENIC
F	MAA	IP	9, 11, 13, OR 15	2.50		RAT	3	TERATOGENIC, FETOTOXIC, INTRAUTERINE DEATH
F	EGdiME	ORAL	8 TO 11	2.77		MOUSE	50, 80	TERATOGENIC, FETOTOXIC
F	EGdiME	ORAL	8 TO 11	3.88		MOUSE	50, 80	TERATOGENIC, FETOTOXIC
F	EGdiME	ORAL	8 TO 11	5.44		MOUSE	50, 80	TERATOGENIC, FETOTOXIC, INTRAUTERINE DEATH
F	EGdiME	ORAL	7 TO 14	22.20		MOUSE	67	100% INTRAUTERINE DEATH, 26% MATERNAL MORTALITY
F	EGEE	INHAL	8-14 OR 15-21		100	RAT	51, 54, 124	POSTNATAL BEHAVIORAL AND NEUROCHEMICAL ALTERATIONS, DELAYED GESTATION
F	EGEE	INHAL	8-14 OR 15-21		200	RAT	51, 54, 124	NEONATAL DEATH AFTER DG 14-20 EXPOSURE
F	EGEE	INHAL	8-14 OR 15-21		300	RAT	51, 54, 124	INTRAUTERINE AND NEONATAL DEATH AFTER DG14-20 EXPOSURE
F	EGEE	INHAL	8-14 OR 15-21		600	RAT	51, 54, 124	INTRAUTERINE AND NEONATAL DEATH AFTER DG 7-13 AND DG14-20 EXPOSURE
F	EGEE	INHAL	8-14 OR 15-21		900	RAT	51, 54, 124	INTRAUTERINE DEATH AFTER DG7-13 AND 14-20 EXPOSURE
F	EGEE	INHAL	8-14 OR 15-21		1200	RAT	52, 53	INTRAUTERINE DEATH AFTER DG7-13 AND 14-20 EXPOSURE
F	EGEE	INHAL	8-14 OR 15-21		200	RAT	52, 53	ETHANOL INTERACTION WITH POSTNATAL BEHAVIORAL AND NEUROCHEMICAL ALTERATIONS
F	EGEE	INHAL	7 TO 16		10	RAT	13	POSTNATAL BEHAVIORAL AND NEUROCHEMICAL ALTERATIONS
F	EGEE	INHAL	7 TO 16		50	RAT	13	NO TERATOGENIC OR FETOTOXIC EFFECT
F	EGEE	INHAL	7 TO 16		250	RAT	13	SLIGHT FETOTOXICITY
F	EGEE	INHAL	7 TO 19		10	RABBIT	13	FETOTOXIC, SLIGHT MATERNAL TOXICITY
F	EGEE	INHAL	7 TO 19		50	RABBIT	13	NO TERATOGENIC OR FETOTOXIC EFFECT
F	EGEE	INHAL	7 TO 19		175	RABBIT	13	NO TERATOGENIC OR FETOTOXIC EFFECT
F	EGEE	INHAL	1 TO 18		160	RABBIT	1, 21	FETOTOXIC
F	EGEE	INHAL	1 TO 18		615	RABBIT	1, 21	TERATOGENIC, SLIGHT MATERNAL TOXICITY
F	EGEE	INHAL	1 TO 19		200	RAT	1, 21	TERATOGENIC, SEVERE MATERNAL TOXICITY
F	EGEE	INHAL	1 TO 19		765	RAT	1, 21	TERATOGENIC, FETOTOXIC
F	EGEE	ORAL	8-10, 11-13, 14-16	2.22		RAT	93	100% INTRAUTERINE DEATH, SLIGHT MATERNAL TOXICITY
F	EGEE	ORAL	8 TO 16	2.22		RAT	93	STAGE-SPECIFIC TERATOGENICITY, FETOTOXIC, NO MATERNAL TOXICITY
F	EGEE	ORAL	7 TO 14	40.10		MOUSE	67	TERATOGENIC, FETOTOXIC, INTRAUTERINE DEATH, MATERNAL TOXICITY
F	EGEE	ORAL	1 TO 21	0.13		RAT	69	100% INTRAUTERINE DEATH, 10% MATERNAL MORTALITY
F	EGEE	ORAL	1 TO 21	0.26		RAT	69	NO TERATOGENIC OR FETOTOXIC EFFECT
F	EGEE	ORAL	1 TO 21	0.52		RAT	69	NO TERATOGENIC OR FETOTOXIC EFFECT, EMBRYOTOXIC
F	EGEE	ORAL	1 TO 21	1.03		RAT	69	TERATOGENIC, FETOTOXIC

GLYCOL ETHER MALE REPRODUCTIVE TOXICITY

SEX	GLYCOL	ROUTE	DURATION	MMOL/KG/DAY	PPM	SPECIES	REFERENCE	OBSERVATIONS
M	EG	ORAL	5 WEEKS	8.05		MOUSE	48, 50	NO EFFECT ON TESTIS WEIGHT, RBC, OR WBC
M	EG	ORAL	5 WEEKS	16.11		MOUSE	48, 50	NO EFFECT ON TESTIS WEIGHT, RBC, OR WBC
M	EG	ORAL	5 WEEKS	32.22		MOUSE	48, 50	NO EFFECT ON TESTIS WEIGHT, RBC, OR WBC
M	EG	ORAL	5 WEEKS	64.43		MOUSE	48, 50	NO EFFECT ON TESTIS WEIGHT, RBC, OR WBC
M	EGA	ORAL	5 WEEKS	4.80		MOUSE	50	NO EFFECT ON TESTIS WEIGHT, RBC, OR WBC
M	EGA	ORAL	5 WEEKS	9.60		MOUSE	50	NO EFFECT ON TESTIS WEIGHT, RBC, OR WBC
M	EGA	ORAL	5 WEEKS	19.21		MOUSE	50	NO EFFECT ON TESTIS WEIGHT, RBC, OR WBC
M	EGdiA	ORAL	5 WEEKS	3.42		MOUSE	50	NO EFFECT ON TESTIS WEIGHT, RBC, OR WBC
M	EGdiA	ORAL	5 WEEKS	6.84		MOUSE	50	NO EFFECT ON TESTIS WEIGHT, RBC, OR WBC
M	EGdiA	ORAL	5 WEEKS	13.68		MOUSE	50	NO EFFECT ON TESTIS WEIGHT, RBC, OR WBC
M	EGME	INHAL	3 HOURS		7500	RAT	12	REDUCED TESTIS WEIGHT
M	EGME	INHAL	4 HOURS		150	RAT	12	NO EFFECT ON TESTIS WEIGHT
M	EGME	INHAL	4 HOURS		300	RAT	12	NO EFFECT ON TESTIS WEIGHT
M	EGME	INHAL	4 HOURS		625	RAT	12	NO EFFECT ON TESTIS WEIGHT, SOME HISTOLOGIC CHANGE
M	EGME	INHAL	4 HOURS		1250	RAT	12	REDUCED TESTIS WEIGHT, ALTERED TESTIS HISTOLOGY
M	EGME	INHAL	4 HOURS		2500	RAT	12	REDUCED TESTIS WEIGHT, ALTERED TESTIS HISTOLOGY, MILD TOXICITY
M	EGME	INHAL	4 HOURS		5000	RAT	12	REDUCED TESTIS WEIGHT, ALTERED TESTIS HISTOLOGY, MILD TOXICITY
M	EGME	INHAL	4 HOURS		1000	RAT	12	REDUCED TESTIS WEIGHT, ALTERED TESTIS HISTOLOGY, NO RECOVERY 19 DAYS POST-TREATMENT
M	EGME	INHAL	4 HOURS		2500	RAT	12	REDUCED TESTIS WEIGHT, ALTERED TESTIS HISTOLOGY, NO RECOVERY 19 DAYS POST-TREATMENT
M	EGME	INHAL	5 DAYS		500	RAT	38, 39	NO EFFECT ON FERTILITY
M	EGME	INHAL	5 DAYS		25	RAT	38, 39	TEMPORARY INFERTILITY
M	EGME	INHAL	5 DAYS		25	MOUSE	38, 39	NO EFFECT ON SPERM HEAD MORPHOLOGY
M	EGME	INHAL	5 DAYS		500	MOUSE	38, 39	INCREASED ABNORMAL SPERM HEAD MORPHOLOGY
M	EGME	INHAL	9 DAYS		300	RAT	41	NO EFFECT ON TESTIS WEIGHT, REDUCED WBC
M	EGME	INHAL	9 DAYS		1000	RAT	41	NO EFFECT ON TESTIS WEIGHT, REDUCED WBC, REDUCED THYMUS WEIGHT
M	EGME	INHAL	9 DAYS		100	MOUSE	41	NO EFFECT ON TESTIS WEIGHT
M	EGME	INHAL	9 DAYS		300	MOUSE	41	NO EFFECT ON TESTIS WEIGHT
M	EGME	INHAL	9 DAYS		1000	MOUSE	41	REDUCED TESTIS WEIGHT; REDUCED THYMUS WEIGHT
M	EGME	INHAL	10 DAYS		100	RAT	11, 12	ALTERED TESTIS AND THYMUS WEIGHT, HISTOLOGY, MILD TOXICITY, REDUCED RBC, WBC
M	EGME	INHAL	10 DAYS		300	RAT	11, 12	REDUCED FERTILITY, ALTERED NEUROCHEMICALS IN PUPS SIRED BY EXPOSED MALES
M	EGME	INHAL	6 WEEKS		55	RAT	43, 45	NO EFFECT ON TESTIS WEIGHT, NO HISTOLOGIC CHANGE
M	EGME	INHAL	13 WEEKS		100	RAT	43, 45	REDUCED BODY WEIGHT, TESTIS WEIGHT, THYMUS WEIGHT, WBC, SEVERE HISTOLOGIC CHANGE
M	EGME	INHAL	13 WEEKS		30	RAT	43, 45	NO EFFECT ON TESTIS WEIGHT, NO HISTOLOGIC CHANGE
M	EGME	INHAL	13 WEEKS		300	RAT	43, 45	REDUCED BODY WEIGHT, TESTIS WEIGHT, THYMUS WEIGHT, WBC, SEVERE HISTOLOGIC CHANGE
M	EGME	INHAL	13 WEEKS		30	RABBIT	43, 45	NO EFFECT ON TESTIS WEIGHT, SLIGHT TO NO HISTOLOGIC CHANGE
M	EGME	INHAL	13 WEEKS		100	RABBIT	43, 45	NO EFFECT ON TESTIS WEIGHT, MODERATE HISTOLOGIC CHANGE
M	EGME	INHAL	13 WEEKS		300	RABBIT	43, 45	REDUCED TESTIS WEIGHT, THYMUS WEIGHT, WBC, RBC, SEVERE HISTOLOGIC CHANGE
M	EGME	INHAL	13 WEEKS		30	RAT	62, 116	NO EFFECT ON TESTIS WEIGHT OR FERTILITY
M	EGME	INHAL	13 WEEKS		100	RAT	62, 116	NO EFFECT ON TESTIS WEIGHT OR FERTILITY
M	EGME	INHAL	13 WEEKS		300	RAT	62, 116	REDUCED TESTIS WEIGHT AND FERTILITY, ALTERED HISTOLOGY
M	EGME	ORAL	1 DAY	3.29		RAT	94	SPERMATOCYTE DEGENERATION
M	EGME	ORAL	2 DAYS	6.57		RAT	94	REDUCED TESTIS WEIGHT, HISTOLOGY CHANGES AFTER 1 DOSE
M	EGME	ORAL	5 DAYS	6.57		RAT	15, 16	NO EFFECT ON FERTILITY
M	EGME	ORAL	5 DAYS	1.31		RAT	95, 121	INFERTILITY, ABNORMAL SPERM HEAD MORPHOLOGY, REDUCED SPERM NUMBERS
M	EGME	ORAL	5 DAYS	2.63		RAT	95, 121	INFERTILITY, ABNORMAL SPERM HEAD MORPHOLOGY, REDUCED SPERM NUMBERS
M	EGME	ORAL	7 DAYS	3.29		RAT	15, 16	REDUCED TESTIS WEIGHT, HISTOLOGY CHANGES AFTER 1 DOSE
M	EGME	ORAL	1-10 DAYS	1.97		RAT	5, 115	SPERMATOCYTE DEGENERATION AFTER 1 DOSE, NO EFFECT ON ABP OR FLUID PRODUCTION
M	EGME	ORAL	1-11 DAYS	0.66		RAT	99	NO SPERMATOGENIC ABNORMALITIES AFTER 11 DAYS DOSING
M	EGME	ORAL	1-11 DAYS	1.31		RAT	99	FIRST DEGENERATIVE CHANGES AFTER 2 DOSES, WIDESPREAD DEGENERATION OF SPERMATOCYTES

Figure 1. Exposure data base on the reproductive toxicity of glycol ethers.

REFERENCE: Number refers to the reference in an accompanying list which gives the authors and journal citation.

OBSERVATIONS: A brief summarization of the significant toxic signs (or the absence of them) at each dose or exposure concentration.

REFERENCES

Andrew, F. D. and Hardin, B. D., 1984, *Environ Health Perspect*, 57, 13.
Bachmann, E. and Golberg, L., 1971, *Food Cosmet Toxicol* 9, 39.
Brown, N. A., Holt, D. and Webb, M., 1984, *Toxicol Lett*, 22, 93.
Campbell, J., Holt, D. and Webb, M., 1984, *J Appl Toxicol*, 4, 35.
Chapin, R. E. and Lamb, J. C. IV, 1984, *Environ Health Perspect*, 57, 219.
Cheever, K. L., Plotnick, H. B., et al., 1984, *Environ Health Perspect*, 57, 241.
Clapp, D. E. Zaebst, D. D. and Herrick, R. F., 1984, *Environ Health Perspect*, 57, 91.
Cook, R. R., VanPeenen, P. F. D., et al., 1982, *Arch Environ Health*, 37, 346.
Cullen, M. R., Rado, T., et al., 1983, *Arch Environ Health*, 38, 347.
Dodd, D. E., Snellings, et al., 1983, *Toxicol Appl Pharmacol*, 68, 405.
Doe, J. E. Samuels, D. M., et al., 1983, *Toxicol Appl Pharmacol*, 69, 43.
Doe, J. E., 1984, *Environ Health Perspect*, 57, 199.
Doe, J. E., 1984, *Environ Health Perspect*, 57, 33.
Donley, D. E., 1936, *J Ind Hyg Toxicol*, 18, 571.
Foster, P. M. D., Creasy, D. M., et al., 1983, *Toxicol Appl Pharmacol*, 69, 385.
Foster, P. M. D., Creasy, D. M., et al., 1984, *Environ Health Perspect*, 57, 207.
Gebhardt, D. O. E., 1968, *Teratology*, 1, 153.
Goldberg, M. E., Johnson, H. E., et al., 1964, *Am Ind Hyg Assoc J*, 25, 369.
Greenburg, L., Mayers, M. R. et al., 1938, *J Ind Hyg Toxicol*, 20, 134.
Hanzlik, P. J., Lawrence, W. S., et al., 1947, *J Ind Hyg Toxicol*, 29, 325.
Hardin, B. D., Bond, G. P., et al., 1981, *Scand J Work Environ Health*, 7(Suppl 4), 66.
Hardin, B. D., Niemeier, R. W., et al., 1982, *Drug Chem Toxicol*, 5, 277.
Hardin, B. D., 1983, *Toxicology*, 27, 91.
Hardin, B. D., Goad, P. T. and Burg, J. R., 1984, *Environ Health Perspect*, 57, 69.
Hutson, D. H. L. and Pickering, B. A., 1971, *Xenobiotica*, 1, 105.
Johnson, E. M., Gabel, B. E. G. and Larson, J., 1984, *Environ Health Perspect*, 57, 135.
Jonsson, A.-K., and Steen, G., 1978, *Acta Pharmocol Toxicol*, 42, 354.
Jonsson, A.-K., Pederson, J. and Steen, G., 1982, *Acta Pharmocol Toxicol*, 50, 358.

CHAPTER 13

COMPUTER-AIDED DESIGN (CAD) APPLICATIONS IN INDUSTRIAL HYGIENE

DOAN J. HANSEN

Brookhaven National Laboratory, Safety and Environmental Protection, Upton, New York

Advances in semiconductor technology have prompted widespread utilization of microcomputers in practically all industries. Tools that were once in the hands of professionals in only a few specialized technologies, because of their high costs or their complexity of operation, are now becoming commonly used by other professionals for new applications. Architects and design engineers have benefited from computer-aided design (CAD) for years. CAD systems can draw a variety of shapes and forms, such as buildings or automobiles, and display them on a computer screen in two or three dimensions. Use of CAD systems offers advantages over traditional methods of design. Now, with computer graphics becoming less expensive and offering greater sophistication, CAD capabilities are available to professionals in industrial hygiene as well. For example, designs made on CAD systems are easily altered or corrected, and changes in scale or proportion may be made for prototypes or patterns.

The microcomputer revolution in many respects was initiated, and has continued, because certain arduous tasks may be accomplished much faster with a microcomputer than without (such as typing or statistical calculations). But CAD systems are somewhat different. For most industrial hygienists, the application of skills and expertise related to design has never been a part of industrial hygiene practice. This has been, perhaps in part, because there was no vehicle for applying engineering design techniques. Therefore, the advent of inexpensive, sophisticated CAD systems for microcomputers offers to industrial hygienists a new methodology as well as a state-of-the-art way to apply it.

Computer-aided design is a computer approach to design. The computer becomes the instrument that the designer or engineer uses to transfer his or her ideas. But the computer is not just a replacement for the drafting table; in fact, the design is enhanced by the use of the computer. In a CAD system, the computer is a tool that helps free the designer to concentrate on designing rather than on additional tasks involved with drawing and measuring.

Table I contains the names of three microcomputer CAD packages with supplier addresses. Examples of CAD applications are illustrated in Figures 1–3.

TABLE I. Sources of CAD Software

Software Name	Supplier
AutoCad	Autodisk, Inc., 150 Shoreline Highway, Bldg. B, Mill Valley, CA 94941, 415/331-0356
Generic CADD	Generic Software, Inc., 13205 N.E. 40th, Bellevue, WA 98005, 800/228-3601
CADKEY	Micro Control Systems, Inc., 27 Hartford Turnpike, Vernon, CT 06066, 203/647-0220

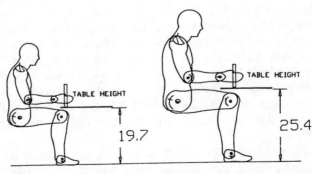

Figure 1. CAD applications in ergonomics. Using CAD, it is possible to compare the optimal workplace requirements of individuals of different statures. Left, fifth percentile female; right, ninety-fifth percentile male. Source: Armstrong et al.(1)

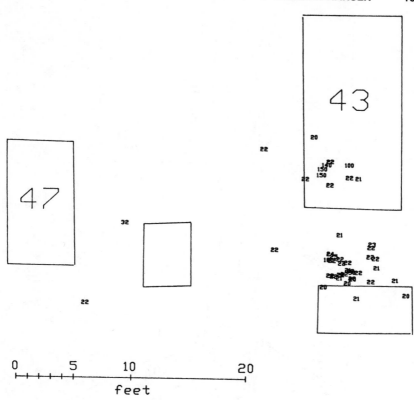

Figure 2. CAD applications in industrial hygiene. Continuous area monitors may simplify the measurement of worker exposures. Using a photoionization detector and a microcomputer equipped with CAD and a digitizor, it is possible to use CAD to plot a worker's instantaneous exposures at their correct locations in the workplace. Source: Hansen et al.(2)

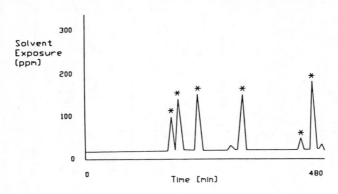
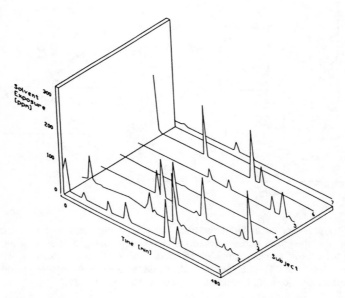

Figure 3. CAD applications in industrial hygiene. Top: one subject's exposure profile is displayed in a conventional X-Y plot. Bottom: using CAD, the same subject's exposure profile can be displayed simultaneously, in three dimensions, along with those of the other workers in the same workplace. Source: Hansen and Whitehead.(3)

REFERENCES

1. Armstrong, T. J., R. G. Radwin, D. J. Hansen, and K. A. Kennedy. 1986. Evaluation and Design of Jobs for Control of Cumulative Trauma Disorders. *Human Factors* (in press).
2. Hansen, D. J., R. G. Radwin, and T. J. Armstrong. 1986. Computer-Aided Design (CAD) in Industrial Hygiene. *Computers in Health and Safety*, K. S. Cohen, Ed. Am. Ind. Hyg. Assoc., Akron, OH.
3. Hansen, D. J. and L. W. Whitehead. 1986. Report on Exposure Group Determination Using Simple Work Measurement Methods (in press).

CHAPTER **14**

RESEARCHING TOXIC SUBSTANCES INFORMATION THROUGH THE USE OF INTERRELATIONAL DATA BASES

MITCHELL BRATHWAITE

Amalgamated Clothing and Textile Workers Union, New York, New York

INTRODUCTION

Burdensome comparative searches for toxic substances information may be alleviated by interrelating data bases. The use of interrelational data bases will be demonstrated utilizing a dBASE II program. Employer chemical inventory information will be manipulated to yield a new data base. The new data base will be reported in a concise, worker-oriented format to produce a chemical hazard summary of the company's inventory.

THE SOFTWARE

A microsoft-Disk Operating System (MS-DOS) program will be used to run the microcomputer. The MS-DOS operating system ties the components of your personal computer (such as the keyboard, monitor, disk drives, and printer) together as a system. The operating system also controls the use and execution of application programs. The application program used in this demonstration is dBASE II; it allows the user to manipulate voluminous amounts of information in a manner which is easily understandable.

TABLE I. CHEMDATA

STRUCTURE FOR FILE: F:CHEMDATA.DBF
NUMBER OF RECORDS: 00046
DATE OF LAST UPDATE: 00/00/00
PRIMARY USE DATA BASE

FLD	NAME	TYPE	WIDTH	DEC
001	CHEM:NAME	C	050	
002	CAS:NUM	C	011	
003	RTECS:NUM	C	009	
004	OSHA:STAND	C	050	
005	RECOMSTAND	C	150	
006	HAZARDS	C	254	
007	SOURCE	C	025	
008	NECPRECAUT	C	254	
** TOTAL **			00804	

THE HARDWARE

The hardware utilized in this exhibit is an IBM-PC with 256K of memory; it is also a dual floppy unit. In addition, there are two monochrome monitors in place and a dot matrix printer. (The second monitor is only necessary for display purposes.)

THE DEMONSTRATION

Two established data bases, CHEMDATA (Table I) and COMPFIL2 (Table II) will be combined to develop a new data base. These data bases contain several fields of information. In order to interrelate the data bases, it is necessary to find similarities between the two groups of information. A close review of the file structures of CHEMDATA and COMPFIL2 would reveal that there are fields in each which are the same, such as CHEM:NAME, CAS:NUM and RTECS:NUM.

With CHEMDATA selected as the primary data base and COMPFIL2 selected as the secondary data base, the JOIN command can be used to combine these two groups of information. The new data base is called NEWDB. For this demonstration the data bases will be interrelated using the key field CHEM:NAME. We can also define the fields that we want on the new data base. The command to perform these tasks is:

    JOIN TO NEWDB FOR CHEM:NAME = S.CHEM:NAME, FIELD CHEM:NAME;
    S.CAS:NUM, S.RTECS:NUM, HAZARDS, TRADENAME, OSHA:STAND;
    NECPRECAUT.

TABLE II. COMPFIL2

STRUCTURE FOR FILE: F:COMPFIL2.DBF
NUMBER OF RECORDS: 00124
DATE OF LAST UPDATE: 00/00/00
SECONDARY USE DATA BASE

FLD	NAME	TYPE	WIDTH	DEC
001	CODE	C	003	
002	TRADENAME	C	100	
003	CHEM:NAME	C	050	
004	DEPARTMENT	C	010	
005	DYECATEGOR	C	050	
006	C:INDEXNUM	C	006	
007	CAS:NUM	C	011	
008	RTECS:NUM	C	009	
** TOTAL **			00240	

TABLE III. NEWDB

STRUCTURE FOR FILE: F:NEWDB.DBF
NUMBER OF RECORDS: 00017
DATE OF LAST UPDATE: 00/00/00
PRIMARY USE DATA BASE

FLD	NAME	TYPE	WIDTH	DEC
001	CHEM:NAME	C	050	
002	TRADENAME	C	100	
003	CAS:NUM	C	011	
004	RTECS:NUM	C	009	
005	SOURCE	C	050	
006	OSHA:STAND	C	050	
007	RECOMSTAND	C	150	
008	HAZARDS	C	254	
009	NECPRECAUT	C	254	
** TOTAL **			00929	

The file structure of NEWDB is displayed in Table III.

Through the use of the REPORT FORM command contained in the dBASE II program, the new data base can be displayed in an easily understandable format (Table IV).

TABLE IV. Amalgamated Clothing and Textile Workers Union Department of Occupational Safety and Health Chemical Hazard Summary

Chemical Name	Trade Names	OSHA Standard a/	Recommended Standard a/b/	Hazards	Necessary Precautions
Antimony oxide [Stibine]	5610 Fire retardant adhesive	0.5 mg/m³	Same	It is an irritant of the nose, throat, eyes and skin. Injury may occur to moist areas of the skin and nose exposed to antimony compounds. Other symptoms include headaches, tightness in chest, vomiting and diarrhea.	Use local exhaust ventilation to control antimony oxide dust levels. A combination of protective clothing, gloves and goggles should be used. Eye wash, washing and showering facilities.
Biphenyl [Diphenyl, phenylbenzene]	Carolid RNO	1.0 mg/m³ (0.2 ppm)	ACGIH: 1.5 mg/m³	Exposure to biphenyl may cause irritation of the eyes and throat. Repeated exposure may cause headache, nausea, stomach pain, fatigue, numbness and aching limbs, and liver damage.	Use ventilation to remove biphenyl dust or vapors. To prevent exposure to the skin and eyes, use protective clothing, gloves and goggles.
Cadmium (dust)	Zinc dust	0.2 mg/m³	NIOSH: 0.04 mg/m³ ACGIH: 0.05 mg/m³	It is a lung irritant. Symptoms of poisoning are fluid accumulation in the lung, chest tightening, headaches, muscle aches, vomiting and diarrhea. Symptoms may be delayed 4 and 10 hrs. IT CAUSES CANCER IN ANIMALS, AND POTENTIALLY CAUSES CANCER IN HUMANS.	To control cadmium airborne levels use local exhaust and general ventilation. Protective clothing should be used and changed periodically. Use dust respirators for added protection and emergencies. Workers should shower after each shift before changing.
Chromates	#140 chromate indicator (contains water soluble potassium chromate)	0.1 mg/m³ (ceiling)	NIOSH: 0.025 mg/m³ ACGIH: 0.05 mg/m³ for water soluble chromium (VI) NIOSH: 0.001 mg/m³ ACGIH: 0.05 mg/m³ for water insoluble chromium (VI).	Chromate dust may cause severe soreness of the nose, throat, bronchial tubes and lungs. If swallowed, chromates may cause stomach and kidney problems. CERTAIN FORMS OF CHROMATES HAVE BEEN KNOWN TO CAUSE CANCER IN HUMANS.	Impervious protective clothing, gloves and footwear should be worn to prevent body contact with chromic acid or chromates. Safety goggles or face shields should be worn in areas where spills and splashes may occur. Use ventilation to remove dust.

CHAPTER 15

METABOLIC ENERGY EXPENDITURE PREDICTION AND STATIC STRENGTH PREDICTION PROGRAM MODELS

CHARLES WOOLEY and RANDY RABOURN
University of Michigan, Center for Ergonomics, Ann Arbor, Michigan

INDIRECT METABOLIC ENERGY EXPENDITURE PREDICTION

Indirect methods of measuring metabolic energy expenditure are by their nature predictions. The accuracy of the prediction depends upon the relationship between the factors being monitored and the actual energy being expended. The prediction model, to be examined below, is based on the assumption that a job can be divided into tasks (activity elements) and that the energy expenditure of the job can be found by summing the energy of these tasks. The energy expenditures of the relatively simple tasks are calculated using mathematical equations derived from empirical data. The accuracy of this prediction procedure depends upon: (1) the completeness and accuracy of the division of the job into tasks, (2) the availability of an equation to precisely describe the task which has been identified, and (3) the accuracy of the task equation itself.

The following is a brief description of the model. The general equation is:

$$\overline{\dot{E}}_{job} = \frac{\sum_{L=1}^{nl} E_{pos_i} \times t_i + \sum_{L=1}^{n} \Delta E_{task_i}}{T}$$

where: $\bar{E}_{job}$ = average metabolic expenditure rate of job

$\dot{E}_{pos_i}$ = metabolic energy expenditure rate of maintaining i^{th} posture

t_i = time duration of ith posture

ΔE_{task_i} = net increase in metabolic energy expenditure of ith task due to external work

T = time duration of the job

The model accommodates three different posture rates ($\dot{E}_{pos_i}$). The computer version of the model requires the analyst to input the percentage of time that the worker spends in each of these postures while performing his or her job.

Posture Tasks (or Rates)

The three posture tasks are:

Standing
Standing bent
Sitting

Incremental Tasks

In addition to the posture tasks, the model accommodates 26 different incremental tasks (ΔE_{task_i}). They are grouped into several major divisions. The computer version of the model requires the input of certain parameters for each of the following incremental tasks.

Lifts:
Stoop—back bent and knees straight, up to 32"
Squat—back straight and knees bent, up to 32"
Semisquat—back and knees both partially bent
One hand—one hand used, up to 32"
Arm—back and knees straight, above 32"

Lowers:
Stoop—back bent and knees straight, below 32"
Squat—back straight and knees bent, below 32"
Semisquat—back and knees both partially bent
Arm—back and knees straight, above 32"

Walking:
> Walking on level or inclined surface

Carrying:
> Loads held at arm's length at sides (in one or both hands)
> Loads held against waist or thighs

Holding:
> Loads held at arm's length at sides (in one or both hands)
> Loads held against thighs or waist

Pushing/pulling:
> Horizontal, forward direction
> At any height from floor

Hand work:
> Light—writing, hand knitting, mechanical accounting
> Heavy—typing, gear assembly, working at lathe

Lateral arm work 180 degrees:
> Lateral movement of arms in horizontal plane
> Bench height, standing, legs moving
> One or both hands
> Example—pick load up from bench in front of body, step back, turn 180 degrees, and place load on second bench.

Lateral arm work 90 degrees:
> Lateral movement of arms in horizontal plane
> Bench height, feet stationary
> One hand or both
> Sitting or standing
> Example—pick up load from bench in front of body, twist 90 degrees, and place load on second bench.

Horizontal arm work:
> Loads moved horizontally forward
> Standing or sitting
> One hand or both
> Example—pick up load from bench 2 inches in front of body and move it 14 inches out from body.

General arm work:

    Light work — loads less than 5 pounds, e.g., filing metal, planing wood, etc.

    Heavy work — loads or forces of more than 5 pounds or fast movements with little rest between exertions, e.g., hammering in nails, shoe making, upholstering, raking garden, etc.

The following is an example job used to demonstrate the use of the model.

EXAMPLE JOB

This is an imaginary job performed at The University of Michigan Printing Services Facility. (It has been said that many jobs there are imaginary.) The worker loads boxes (cases) of paper into a box labeling machine. The boxes are stacked three high and are located on the floor at a distance of ten feet from the machine. They weigh 20 pounds each and measure 12 inches on each side. The worker lifts each of the boxes and carries it to the machine. He then pushes the box into the machine and walks back to get the next box. Twice a day the machine must be loaded with ink and checked for proper operation. At the end of the shift it must be cleaned.

Figure 1 illustrates the input form used for the energy expenditure model. A listing of the posture tasks, the incremental tasks, and the parameters for this job can be found in Table I.

DESCRIPTION OF EXPENDITURE PREDICTION PROGRAM

The Metabolic Energy Expenditure Prediction Program is a microcomputer-based model which can estimate the energy requirement of a wide variety of manual materials handling jobs. The model is based on the assumption that a job can be divided into simple tasks (activity elements) and that the average metabolic energy expenditure rate of the job can be predicted by knowing the energy expenditures of the simple tasks and the time duration of the job.

Model Inputs

- Subject gender and body weight
- List of activity elements, e.g., lift, push, carry, etc.

PLANT: U of M Printing **DATE:** 5/20/84 **TIME:** 2 PM **ANALYST:** CBW
JOB TITLE: Machine Loader **JOB NUMBER:** 1001
WORKER ID: 2634 **HEIGHT:** 210 **GENDER:** Male
DURATION OF CYCLE: 1 MIN **FREQUENCY OF CYCLE:** 60/HOUR
COMMENTS: #2 Load three box stack into machine

TASK NUMBER	TASK DESCRIPTION	FREQ./CYCLE	FORCE POUNDS	INITIAL POSITION¹	FINAL POSITION²	TIME MIN.	DISTANCE FEET	SLOPE %
201	Lift Top Box, Stoop	1	20	24"	34"			
202	Lift Mid Box, Semi-Squat	1	20	12"	34"			
203	Lift Low Box, Stoop	1	20	0"	34"			
210	Carry, Waist	3	20			.06	10'	0
220	Lower To Table	3	20	34"	28"			
230	Push	3	10	38"	18"			
240	Walk	3				.05	10'	0
	Standing	80%						
	Standing Bent	20%						

(1) - for push/pull vert. height of hands   (2) - for push/pull horz. distance

Figure 1. Data input form for energy expenditure model.

TABLE I. Metabolic Expenditure Prediction Model

Job: Machine loader
Number: 1001
Job frequency: 60/Hour
Subject ID: 2634
Subject: Male, 210 pounds
Analyst: CBW
Date: 5/20/84

Tasks:

Num	Freq	Type	Location Initial	Location Final	Load LB	Grade %	Time Min	Incremental Energy Kcal
201	1	Stoop lift pick up tool box	24	34	20			0.12
202	1	Semisquat lift pick up middle box	12	34	20			0.33
203	1	Stoop lift pick up lowest box	0	34	20			0.43
210	3	Carry at waist carry box to table	0	10	20	0	0.060	0.67
220	3	Stoop lower place box on table	34	28	20			0.15
230	3	Push/pull @ 38 in. push box toward machine	0	18	10			0.24
240	3	Walk return to stack of boxes	0	10		0	0.050	0.46
		Total incremental energy						2.40
		Standing energy cost (0.80 min)						1.83
		Standing bent energy cost (0.20 min)						0.54
		Total energy expenditure (Kcal)						4.77
		Job cycle duration (MIN)						1.00
		Energy expenditure rate (Kcal/min)						4.77

- Parameters specific to the activity elements, e.g., frequency, weight of load, distance carried, etc.

Model Outputs

- Listing of all activity elements and parameters including the energy expenditure associated with each element
- Calculation of the total energy expenditure for the job cycle
- Calculation of the total energy expenditure rate in Kcal/minute for the job
- Data can be output to a disk file for recall at a later date

System Requirements

- IBM-PC computer
- 128K random access memory
- 5¼" single-sided floppy disk drive

Availability

IBM version is available which includes program disk and documentation.

DESCRIPTION OF STATIC STRENGTH PREDICTION PROGRAM

The Static Strength Prediction Program is a microcomputer-based, sagittal plane, biomechanical prediction model. It is designed to predict human static strength requirements and low back stress levels. General subject anthropometric, posture, and the load acting on the hands are required input information as shown in Figure 2. The model was designed for manual materials handling job design and evaluation and for instructional purposes.

Model Inputs

- Male or female stature and weight data of 5, 50, or 95th percentile population groups
- Forearm, upper arm, torso, upper leg, and lower leg postures
- Load magnitude and direction acting on the hand(s)

Model Outputs

- Stick figure posture descriptions (graphics monitors)
- Predictions of (1) male and female elbow, shoulder, torso, hip, knee and ankle strength requirements, (2) male and female elbow, shoulder, torso, hip, knee and ankle population strength capabilities, (3) resultant forces at joints, (4) L5/S1 (5th lumbar/1st sacral) disk compression forces, (5) body link lengths, masses, and centers of mass locations, (6) abdominal pressure, and (7) back muscle force

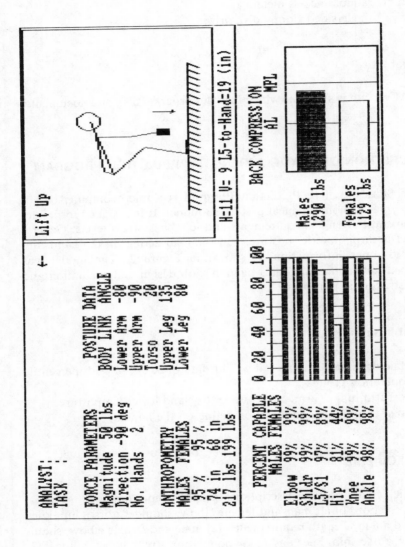

Figure 2. Example of Static Strength Predition program capabilities.

System Requirements

IBM

- PC Computer or clone
- 128K random access memory
- IBM Graphics capability
- Epson dot matrix printer (optional)
- 5¼" double-sided floppy disk drive
- DOS 2.0 or greater

Macintosh

- 512K
- Microsoft basic
- Apple printer

Availability

Version 3.8c is available for the IBM and Version 3.0 for the Macintosh. The program includes a disk and documentation/tutorial instructions.

INFORMATIONAL CONTACT

Hira Herrington, The University of Michigan, Center for Ergonomics, IOE Building, 1205 Beal Avenue, Ann Arbor, MI 48109-2117; (313)763-5773.

CHAPTER 16

EXAMPLE PROBLEM—
COMPUTER-AIDED VENTILATION DESIGN

DAVID E. CLAPP, PhD, PE

National Institute for Occupational Safety and Health, Cincinnati, Ohio

The design of a ventilation system requires use of a good reference guide or textbook. For most needs, the *Industrial Ventilation—A Manual of Recommended Practice (Ventilation Manual)*, published by the American Conference of Governmental Industrial Hygienists should be adequate. It is necessary to have a basic understanding of ventilation design before using this program. The program will then aid the user and prompt him in proceeding through the design process.

The design methodology used in the program is the "equivalent foot method." In this method, all elements of the system (e.g., fittings) are converted to equivalent feet of straight duct, hence simplifying the analysis. It is the opinion of the authors that the equivalent foot method generates a superior system over designs produced by the "blast gate method." Readers interested in further discussion of the blast gate method should contact the authors.

The design of a ventilation system begins with a schematic drawing. Such a drawing is shown in Figures 1 and 2. The system displayed will be used as an illustrative example in this user's manual. Note that hood selections have already been made and details extracted from the *Ventilation Manual*. For this design, the details are summarized in Tables I and II.

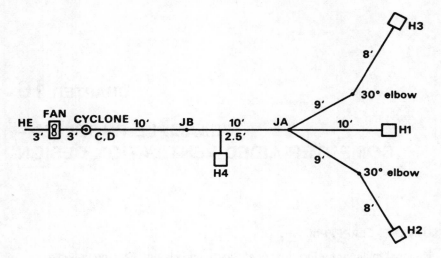

Figure 1. Top view of ventilation system.

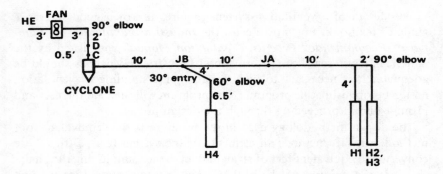

Figure 2. Side view of ventilation system.

TABLE I. Hood Design Data

Hood	Description	VS Print	Min. Exhaust
1	3 × 2 ft welding bench	416	1050
2	Metal cutting bandsaw	418	800
3	Metal cutting bandsaw	418	800
4	Grinder wheel hood	411	500

TABLE II. System Details

Branch	Minimum CFM Required	Length (ft)	Elbows	Entries
H1–JA	1050	14	1–90°	
H2–JA	800	21	1–90°, 1–30°	30°
H3–JA	800	21	1–90°, 1–30°	30°
A–JB	2650	10		
H4–JB	500	13	1–90°, 1–60°	30°
B–C	3150	10.5		
C–D	3150	Expansion	30° taper	
D–FAN	3150	2	1–90°	
Fan HE	3150	3		
Cyclone	–	1 in. W.G. pressure drop		

The reader should carefully notice the labeling of the system in Figure 1. Careful labeling is important in this program since the program distinguishes between system elements by the label. The labeling convention may be summarized in the following rules:

1. All points on the schematic must be labeled with one or two letters (e.g., C or JB) or a letter and a number (e.g., H1 or H4). No more than TWO letters (or letter and number) are permitted!
2. All points on the schematic which are openings to the atmosphere MUST be labeled beginning with an "H" (e.g., H1 or HZ).
3. All junctions of two or more branches MUST be labeled beginning with a "J" (e.g., JA).
4. The position of the fan in the schematic MUST be labeled beginning with a "F" (e.g., FN). Note, it is important to remember that labeling the fan location as FAN is illegal since it exceeds two letters! This convention differs from the Apple II version of this program.
5. Internally, the program will relabel all junctions with a single letter (e.g., from JA to A) AFTER the junction is balanced; thus, the user will know at a glance if a junction has been balanced. However, the user must remember to always refer to that point with the single letter (not the original label) after the junction has been balanced. The labeling conventions will become clearer to the reader as the description of the example problem unfolds.

The operation of the ventilation design program begins with loading the program into the computer. The program should begin with the question:

Is this a new system?

This question should be answered yes or no depending if the user is working with an old system file or not. If the answer is NO, the program will simply display the menu shown below. The user then may list the file by using the proper option in the menu (option 4 or 5), or begin working with the design as required.

The function of each of the options is reasonably self-explanatory; however, each of the options will be demonstrated in the sample design in this manual. An option is selected by simply entering the number of the option in response to the "prompt" which appears at the end of the menu. The menu as it appears on the computer screen is as follows:

    1 — Determine the pressure losses in a branch
    2 — Balance all branches at a junction
    3 — Determine a fan's static pressure requirements
    4 — List of branches and their losses
    5 — List of the branches' velocities, total losses, design Q's and duct diameters
    6 — Print branch data on printer
    7 — Determine the static pressure losses or gains through an expansion or contraction
    8 — Calculate the pressures to a point
    9 — Calculate velocity of air for a certain duct size
  10 — Clear ventdata
  11 — Pack ventdata
  12 — Delete a branch
  13 — Rename a branch
  14 — Duplicate a branch
  15 — Redetermine a branch's SP losses
    0 — Exit
Your choice?

In all entries by the user (such as typing the letter "Y" for yes) the entry must be followed by a carriage return. In this manual, the use of the carriage return after each entry will be assumed and will not be stated each time an entry is described.

Whenever a yes answer is desired for questions in the program, the letter "Y" is not explicitly required (in most cases). The carriage return alone is sufficient to input the yes answer. This feature speeds the input since many of the questions used in the program result in yes answers.

The system design is started by selecting the branch furthest from the fan (or the branch with the greatest losses). This selection is relatively arbitrary and for this analysis, the design will begin with branch H1-JA. The design of this branch is accomplished by selecting option 1 and responding to the questions which appear. For this branch, the following data is input:

    Starting point? H1
    Ending point? JA
    Design Q (CFM) 1050
    Transport velocity (FPM) 2000

This data is directly extracted from the schematic and Tables I and II. Upon input of this data, the computer will respond with a selected duct size (10 in.), duct velocity (1925 FPM), and duct velocity pressure (0.231 in. W.G.). The next series of questions obtain data regarding losses in the branch under design including: entry, duct, and fitting losses. This data is input with responses to questions as follows:

Are there any slots	Y

(*Note*: in answering yes only carriage return is required.)

Slot velocity	1000
Slot entry loss factor (VP)	1.78
Hood entry loss factor (VP)	.25
Are there any duct losses	Y
Length (ft)	14
Are there any fittings	Y
Are there any elbows	Y
No. 90 degree elbows	1
No. 60 degree elbows	
No. 45 degree elbows	
No. 30 degree elbows	
Are there any entry losses	N
Any other pressure losses	

The last question is answered by inputting inches (W.G.) of loss or by inputting a "0" (zero) indicating no other pressure losses. This feature is further illustrated later in this design example.

This exercise of entering the above system elements into the computer will result in a calculated loss in inches of water. For the data above, losses should be: slot (.11), hood (.29), duct (.08), fittings (.08), entries (0). The user will note that other results are displayed, such as equivalent length of certain segments (e.g., equivalent length of elbows). The final result is the loss in inches W.G. of the entire branch; in this case, 0.56. Following the display of this output the computer will ask:

Is this satisfactory?   Y

The yes answer to this question causes the branch and its calculated losses to be stored to disk and the process recycles back to the menu of options. At this point, the user may wish to use option 4 or option 5 (or both) to list the branch just designed. This result may be compared to the output listing in Appendix A of this guide.

The user now will choose to enter another branch, H2-JA. The entry is begun by selecting option 1 and answering questions as above; in this case the data is: Q = 800, velocity = 4000, hood loss factor = 1.75, duct length = 21, one 90- and one 30-degree elbow, and one 30-degree entry. Note that the question "Are there entry losses?" should be answered "Y," but "Are there any slots?" answered "N." Then the computer will ask for hood loss factor, omitting reference to slots. This data should produce a branch with losses of hood (2.8), duct (.93), fittings (.42), entry (.23) for a total loss of 4.43 in. W.G.

By examining the system schematic, it is clear that branch H3-JA is identical to branch H2-JA, hence it is only necessary to duplicate H2-JA. This is accomplished by selecting option 14. The inputs for this option are simple and straightforward and the user responds to the questions listed. The result is the addition of an identical branch H3-JA. The user should verify this result by listing the design with options 4 and 5 and comparing the result to the printouts in Appendix A.

The three branches leading to junction JA are complete and it is time to balance this junction. By selecting option 2, a balance at this junction will be attempted (this option is automatic and the user only indicates the name of the junction to be balanced). The result is that a redesign is necessary since branch H1-JA has a far smaller resistance than the other two branches—too small for the computer to balance (either within 5% or within 20%). This redesign is accomplished by choosing option 15. Here the process is trial and error, redesigning the branch until losses are high enough. An educated guess can get the losses close to a desired value. In this case, try reducing the duct size to "6 inches." Answer the questions identically as in the original design of the branch and examine the resulting loss (3.92 in. W.G.). Clearly, this loss is within 20% and the balance may be accomplished.

Before again attempting to balance junction JA, select option 5 and list branches. Note that the first line, formerly H1-JA, is now zeros and a redesigned branch is shown at the bottom of the list. The "zeroed" branch should be removed with option 11 to "pack the listing." Again, list the branches (option 5) and examine the compacted list.

Now a balance may be attempted for junction JA by selecting option 2. This is an automatic procedure and the computer will list the revised flow rate (Q) based upon the square root of the ratio of the static pressures between the higher loss and lower loss branches (see equation in ACGIH, *Ventilation Manual*). The balance is successful with the redesign of branch H1-JA and the user should list all of the branches to review the data. Notice that the last branch listed is HJ-A. This is the equivalent branch for the three incoming branches. The losses in this branch correspond to the including branch with highest losses, the airflow is the sum of incoming airflows, and the diameter is the square root of the squares of the diameters of the incoming ducts (10.39 in.). It is IMPERATIVE to note that junction JA has been renamed A. Hence, all following calculations will refer to A rather than JA. In this manner, the computer can identify junctions which are balanced.

The designer may now continue from the junction onward into the ventilation system. The next branch to be designed is branch A-JB. (Again, notice the use of "A" versus "JA.") To design this branch, select option 1 and the computer will respond with the design Q to this point (2716 CFM) and transport velocity (4612 FPM). The computer will ask "Transport Velocity?" and the user may either enter a desired transport velocity or simply hit enter without typing a number to use the value of the previous branch. In this case, only 4000 FPM is required, hence the user should enter 4000. If the user simply hits "enter" without typing a number, the computer will assume 4612 FPM and the duct will be sized based on that value. The design of this branch continues by specifying no entries, 10 feet of duct, no fittings or extraneous losses, leading to a total branch loss of 0.21 in. W.G.

At any time, the user may wish to compute losses up to a point in the system. For example, the user may wish to know pressure losses to point JB so that he may judge whether a balance will be possible at that junction. To obtain cumulative losses to a point, select option 8 and name the point. For the point JB, the computer should specify the static pressure as −4.64, velocity pressure as 1.05 and total pressure of −3.59 in. W.G. Thus, the user knows that to balance, H4-JB must have losses within 20% of 4.64 in. W.G.

The next step in the design process is to design the other branch into junction JB, which is H4-JB. In this case, the user should enter a Q of 500 CFM, transport velocity of 4500 FPM, hood entry loss factor of 0.65 (no slots), 13 feet of duct, one 90- and one 60-degree elbow, and an entry of 30 degrees. All of these features should result in a total calculated loss of 4.05 in. W.G.

Junction JB may now be balanced. The user knows that the two branches will balance within 20% due to comparing the results of the last two steps. Thus, by selecting option 2 the new flow is specified of 536 CFM to make up the difference in pressure between the two branches. The user should now display results to this point by option 4, 5, or both and notice the equivalent branch HJ-B which has been created due to the balancing option.

Branch B-C may now be designed. Using option 1, the user enters the transport velocity of 4000 FPM, duct length of 10 ft, and an extraneous loss of 1 in. W.G. (Notice that this is the first time extraneous losses have been input.) There are no fittings or entries in this section of duct. This input should produce a calculated loss of 1.19 in. W.G. The user may now wish to compute cumulative losses to point C (option 8) which should produce a calculated value of –5.85 in. W.G.

Since branch C-D is an expansion, the branch which follows it must be designed first. If the user attempts to design a contraction or expansion without designing the follow-on section first, an error message will be received. The branch which follows C-D is D-FN. (As a reminder, note the fan location is designated with an "F" prefix, and only two characters are allowed.) Branch D-FN is designed by selecting option 1 and entering data for the branch including flow of 3253 CMF, 1500 FPM transport velocity, 5 feet of duct, and one 90-degree elbow. There are no entry or extraneous losses.

After designing D-FN, the expansion may be designed by selecting option 7. The questions asked here include whether the section of duct is an expansion or a contraction, taper angle (30 degrees in this case), and if the expansion is within 60 inches of a disturbance (yes, in this case). The data input should result in a calculated regain of –0.36 in. W.G. The user should list the sections designed, study the data, and compare results to Appendix A.

The last branch in the system is branch FN-HE. The use of option 1, hopefully, should be routine for the user by this point. In this case, the question "transport velocity" should be answered with the enter key only (no number typed in) since the previous branch transport velocity is acceptable. The only loss in this branch is 3 feet of duct, leading to a total branch loss of 0.004 in. W.G.

If an error is ever made in labeling a branch, option 13 may be used to correct the labels. For example, if the previous branch had been labeled FN-E, the label could be easily modified to FN-HE by selecting option 13 and answering the questions which appear. The user may wish to modify at least one branch label to exercise this option.

The fan static pressure requirements may be computed by selecting option 3. All that is required for input is the location of the fan (FN in this case) and the computer will produce the static pressure requirement of 5.41 in. W.G. This data will be routed to the printer at the user's option.

The final step in this exercise is to print the results (using option 6) and compare the printed data to the results in Appendix A. If there are substantial differences, the design should be reaccomplished. Any individual branch which appears questionable can be deleted (option 12) and redesign accomplished (option 1). If necessary, the design may be totally reaccomplished by deleting the system in its entirety (option 10) and starting over with option 1.

A user without a computer can also print final data as a screen display. This data will appear in two sections by choosing option 4 or 5. Each option produces one-half of the output produced in the printer display (option 6).

APPENDIX A

Section 1: Determine the pressure losses in a branch (H1–JA)

ST	ED	Q	VE	VP	DD	SLOT	HOOD	DUCT	ELBW	ENTR	TOTAL
H1	JA	1050	1925.1381	.23105680	10	.1109	.2888	.0802	.0769	0	.556985833

Section 1: Determine the pressure losses in a branch (H2–JA)

ST	ED	Q	VE	VP	DD	SLOT	HOOD	DUCT	ELBW	ENTR	TOTAL
H1	JA	1050	1925.1381	.23105680	10	.1109	.2888	.0802	.0769	0	.556985833
H2	JA	800	4074.3664	1.0349399	6	0	2.846	.9326	.4284	.2260	4.43333989

Section 14: Duplicate a branch (H3–JA duplicates H2–JA)

ST	ED	Q	VE	VP	DD	SLOT	HOOD	DUCT	ELBW	ENTR	TOTAL
H1	JA	1050	1925.1381	.23105680	10	.1109	.2888	.0802	.0769	0	.556985833
H2	JA	800	4074.3664	1.0349399	6	0	2.846	.9326	.4284	.2260	4.43333989
H3	JA	800	4074.3664	1.0349399	6	0	2.846	.9326	.4284	.2260	4.43333989

Section 15: Redetermine a branch's SP losses (H1–JA)

ST	ED	Q	VE	VP	DD	SLOT	HOOD	DUCT	ELBW	ENTR	TOTAL
0	0	0	0	0	0	0	0	0	0	0	0
H2	JA	800	4074.3664	1.0349399	6	0	2.846	.9326	.4284	.2260	4.43333989
H3	JA	800	4074.3664	1.0349399	6	0	2.846	.9326	.4284	.2260	4.43333989
H1	JA	1050	5347.6060	1.7828457	6	.1109	2.228	1.042	.5388	0	3.92082829

DAVID E. CLAPP

Section 11: Pack vent data

ST	ED	Q	VE	VP	DD	SLOT	HOOD	DUCT	ELBW	ENTR	TOTAL
H2	JA	800	4074.3664	1.0349399	6	0	2.846	.9326	.4284	.2260	4.43333989
H3	JA	800	4074.3664	1.0349399	6	0	2.846	.9326	.4284	.2260	4.43333989
H1	JA	1050	5347.6060	1.7828457	6	.1109	2.228	1.042	.5388	0	3.92082829

Section 2: Balance all branches at a junction (Junction A)

ST	ED	Q	VE	VP	DD	SLOT	HOOD	DUCT	ELBW	ENTR	TOTAL
H2	JA	800	4074.3664	1.0349399	6	0	2.846	.9326	.4284	.2260	4.43333989
H3	JA	800	4074.3664	1.0349399	6	0	2.846	.9326	.4284	.2260	4.43333989
H1	JA	1116.5184	5686.3816	2.0158906	6	.1254	2.519	1.178	.6093	0	4.43333989
HJ	A	2716.5184	4611.7050	1.3259220	10.39	0	0	4.433	0	0	4.43333989

Section 1: Determine the pressure losses in a branch (A–JB)

ST	ED	Q	VE	VP	DD	SLOT	HOOD	DUCT	ELBW	ENTR	TOTAL
H2	JA	800	4074.3664	1.0349399	6	0	2.846	.9326	.4284	.2260	4.43333989
H3	JA	800	4074.3664	1.0349399	6	0	2.846	.9326	.4284	.2260	4.43333989
H1	JA	1116.5184	5686.3816	2.0158906	6	.1254	2.519	1.178	.6093	0	4.43333989
HJ	A	2716.5184	4611.7050	1.3259220	10.39	0	0	4.433	0	0	4.43333989
A	JB	2716.5184	4116.2324	1.0563181	11	0	0	.2161	0	0	.216172917

Section 1: Determine the pressure losses in a branch (H4–JB)

ST	ED	Q	VE	VP	DD	SLOT	HOOD	DUCT	ELBW	ENTR	TOTAL
H2	JA	800	4074.3664	1.0349399	6	0	2.846	.9326	.4284	.2260	4.43333989
H3	JA	800	4074.3664	1.0349399	6	0	2.846	.9326	.4284	.2260	4.43333989
H1	JA	1116.5184	5686.3816	2.0158906	6	.1254	2.519	1.178	.6093	0	4.43333989
HJ	A	2716.5184	4611.7050	1.3259220	10.39	0	0	4.433	0	0	4.43333989
A	JB	2716.5184	4116.2324	1.0563181	11	0	0	.2161	0	0	.216172917
H4	JB	500	4527.0738	1.2777036	4.5	0	2.127	1.001	.6565	.2720	4.05784628

Section 2: Balance all branches at a junction (Junction B)

ST	ED	Q	VE	VP	DD	SLOT	HOOD	DUCT	ELBW	ENTR	TOTAL
H2	JA	800	4074.3664	1.0349399	6	0	2.846	.9326	.4284	.2260	4.43333989
H3	JA	800	4074.3664	1.0349399	6	0	2.846	.9326	.4284	.2260	4.43333989
H1	JA	1116.5184	5686.3816	2.0158906	6	.1254	2.519	1.178	.6093	0	4.43333989
HJ	A	2716.5184	4611.7050	1.3259220	10.39	0	0	4.433	0	0	4.43333989
A	JB	2716.5184	4116.2324	1.0563181	11	0	0	.2161	0	0	.216172917
H4	JB	536.48053	4857.374	1.4709504	4.5	0	2.427	1.153	.7557	.3132	4.64951281
HJ	B	3252.9989	4222.4847	1.1115554	11.88	0	0	4.649	0	0	4.64951281

Section 1: Determine the pressure losses in a branch (B–C)

ST	ED	Q	VE	VP	DD	SLOT	HOOD	DUCT	ELBW	ENTR	TOTAL
H2	JA	800	4074.3664	1.0349399	6	0	2.846	.9326	.4284	.2260	4.43333989
H3	JA	800	4074.3664	1.0349399	6	0	2.846	.9326	.4284	.2260	4.43333989
H1	JA	1116.5184	5686.3816	2.0158906	6	.1254	2.519	1.178	.6093	0	4.43333989
HJ	A	2716.5184	4611.7050	1.3259220	10.39	0	0	4.433	0	0	4.43333989
A	JB	2716.5184	4116.2324	1.0563181	11	0	0	.2161	0	0	.216172917
H4	JB	536.48053	4857.374	1.4709504	4.5	0	2.427	1.153	.7557	.3132	4.64951281
HJ	B	3252.9989	4222.4847	1.1115554	11.88	0	0	4.649	0	0	4.64951281
B	C	3252.9989	4141.8468	1.0695055	12	0	0	1.196	0	0	1.19670623

Section 1: Determine the pressure losses in a branch (D–FN)

ST	ED	Q	VE	VP	DD	SLOT	HOOD	DUCT	ELBW	ENTR	TOTAL
H2	JA	800	4074.3664	1.0349399	6	0	2.846	.9326	.4284	.2260	4.43333989
H3	JA	800	4074.3664	1.0349399	6	0	2.846	.9326	.4284	.2260	4.43333989
H1	JA	1116.5184	5686.3816	2.0158906	6	.1254	2.519	1.178	.6093	0	4.43333989
HJ	A	2716.5184	4611.7050	1.3259220	10.39	0	0	4.433	0	0	4.43333989
A	JB	2716.5184	4116.2324	1.0563181	11	0	0	.2161	0	0	.216172917

ST	ED	Q	VE	VP	DD	SLOT	HOOD	DUCT	ELBW	ENTR	TOTAL
H4	JB	536.48053	4857.374	1.4709504	4.5	0	2.427	1.153	.7557	.3132	4.64951281
HJ	B	3252.9989	4222.4847	1.1115554	11.88	0	0	4.649	0	0	4.64951281
B	C	3252.9989	4141.8468	1.0695055	12	0	0	1.196	0	0	1.19670623
D	FN	3253	1491.0653	.13860800	20	0	0	7.E-3	.0470	0	.054604462

Section 7: Determine the static pressure losses or gains through an expansion or contraction (C–D)

ST	ED	Q	VE	VP	DD	SLOT	HOOD	DUCT	ELBW	ENTR	TOTAL
H2	JA	800	4074.3664	1.0349399	6	0	2.846	.9326	.4284	.2260	4.43333989
H3	JA	800	4074.3664	1.0349399	6	0	2.846	.9326	.4284	.2260	4.43333989
H1	JA	1116.5184	5686.3816	2.0158906	6	.1254	2.519	1.178	.6093	0	4.43333989
HJ	A	2716.5184	4611.7050	1.3259220	10.39	0	0	4.433	0	0	4.43333989
A	JB	2716.5184	4116.2324	1.0563181	11	0	0	.2161	0	0	.216172917
H4	JB	536.48053	4857.374	1.4709504	4.5	0	2.427	1.153	.7557	.3132	4.64951281
HJ	B	3252.9989	4222.4847	1.1115554	11.88	0	0	4.649	0	0	4.64951281
B	C	3252.9989	4141.8468	1.0695055	12	0	0	1.196	0	0	1.19670623
D	FN	3253	1491.0653	.13860800	20	0	0	7.E-3	.0470	0	.054604462
C	D	3252.9989	1491.0653	.13860800	20	0	0	-.359	0	0	-.35931725

Section 1: Determine the pressure losses in a branch (FN–HE)

ST	ED	Q	VE	VP	DD	SLOT	HOOD	DUCT	ELBW	ENTR	TOTAL
H2	JA	800	4074.3664	1.0349399	6	0	2.846	.9326	.4284	.2260	4.43333989
H3	JA	800	4074.3664	1.0349399	6	0	2.846	.9326	.4284	.2260	4.43333989
H1	JA	1116.5184	5686.3816	2.0158906	6	.1254	2.519	1.178	.6093	0	4.43333989
HJ	A	2716.5184	4611.7050	1.3259220	10.39	0	0	4.433	0	0	4.43333989
A	JB	2716.5184	4116.2324	1.0563181	11	0	0	.2161	0	0	.216172917
H4	JB	536.48053	4857.374	1.4709504	4.5	0	2.427	1.153	.7557	.3132	4.64951281
HJ	B	3252.9989	4222.4847	1.1115554	11.88	0	0	4.649	0	0	4.64951281
B	C	3252.9989	4141.8468	1.0695055	12	0	0	1.196	0	0	1.19670623
D	FN	3253	1491.0653	.13860800	20	0	0	7.E-3	.0470	0	.054604462

192   MICROCOMPUTER APPLICATIONS IN OH&S

ST	ED	Q	VE	VP	DD	SLOT	HOOD	DUCT	ELBW	ENTR	TOTAL
C	D	3252.9989	1491.0653	.13860800	20	0	0	-.359	0	0	-.35931725
FN	E	3253	1491.0653	.13860800	20	0	0	4.E-3	0	0	4.5E-3

Section 13: Rename a branch (FN–E should be FN–HE)

ST	ED	Q	VE	VP	DD	SLOT	HOOD	DUCT	ELBW	ENTR	TOTAL
H2	JA	800	4074.3664	1.0349399	6	0	2.846	.9326	.4284	.2260	4.43333989
H3	JA	800	4074.3664	1.0349399	6	0	2.846	.9326	.4284	.2260	4.43333989
H1	JA	1116.5184	5686.3816	2.0158906	6	.1254	2.519	1.178	.6093	0	4.43333989
HJ	A	2716.5184	4611.7050	1.3259220	10.39	0	0	4.433	0	0	4.43333989
A	JB	2716.5184	4116.2324	1.0563181	11	0	0	.2161	0	0	.216172917
H4	JB	536.48053	4857.374	1.4709504	4.5	0	2.427	1.153	.7557	.3132	4.64951281
HJ	B	3252.9989	4222.4847	1.1115554	11.88	0	0	4.649	0	0	4.64951281
B	C	3252.9989	4141.8468	1.0695055	12	0	0	1.196	0	0	1.19670623
D	FN	3253	1491.0653	.13860800	20	0	0	7.E-3	.0470	0	.054604462
C	D	3252.9989	1491.0653	.13860800	20	0	0	-.359	0	0	-.35931725
FN	HE	3253	1491.0653	.13860800	20	0	0	4.E-3	0	0	4.5E-3

Fan static pressure difference (W.G) = 5.40744036; fan design Q (CFM) = 3253; intake duct diameter (in.) = 20; output duct diameter (in.) = 20.

CHAPTER 17

USE OF A MICROCOMPUTER SPREADSHEET PROGRAM TO MODEL CARBON MONOXIDE LEVELS AND EFFECTS DUE TO INDOOR CONSTRUCTION WITH PARTIAL GENERAL VENTILATION

L.W. WHITEHEAD

School of Public Health, The University of Texas Health Science Center at Houston, Houston, Texas

The modification or rehabilitation of large industrial facilities often requires heavy construction activities inside buildings. This usually requires combustion-powered equipment with the corresponding risk of exhaust product accumulation, especially carbon monoxide. Two situations of this type were investigated in which heavy construction had occurred previously inside industrial buildings. In the larger case, numerous vehicles and equipment were running at various times in the building, without precise information regarding times and locations. The problem was appropriate for a simple model to estimate the effects of various assumptions regarding equipment and ventilation.

To calculate levels of carbon monoxide, an interactive model was constructed using the Symphony™ spreadsheet program, although most any spreadsheet program should be usable. It allows the user to specify a wide range of variables, including building dimensions and openings, vehicle and equipment numbers and emissions, general ventilation assumptions, natural ventilation, and weather conditions. The model then calculates the carbon monoxide concentrations resulting under both a steady-state model and a build-up model (both based on the generalized, simple mass-balance equations for general ventilation). The user specifies the time period for the build-up model, and the time-weighted average (TWA) of the build-up model is also calculated using the definite integral of the build-up equation. The model also calculates the carboxyhemoglobin (COHb) levels resulting from any specified number of hours exposure to the resultant CO levels using the Peterson-Stewart equation, the Coburn equation (which permits input of work rate and accounts for CO elimination as well as uptake), and a simple model by Ott and Mage that estimates COHb levels resulting from changing CO levels. The spreadsheet program also can generate tables of projections of any one variable given specified values of any other one or two variables. These tables are utilized to examine CO levels resulting from various levels of forced and natural ventilation, as well as the associated carboxyhemoglobin levels. This model has proved to be a very useful application of the power inherent in spreadsheet programs.

The three key panels of the spreadsheet are illustrated. Table I is the main section in which most parameters and all results for any one set of conditions are displayed. Tables II and III, respectively, present the major tabulations of ventilation and carbon monoxide sources which are the key variables entered into the model calculations.

The major problem case demonstrated that CO levels ranging from sub-TLV to over 200 ppm were possible given various levels of ventilation and emissions. A small number of historical sampling data existed and all data were consistently in the range predicted by the model under various assumptions. The problem studied demonstrates that major construction conducted inside even a very large structure can result in excessive levels of carbon monoxide. This should be anticipated and critical situations examined prior to beginning any project requiring such activity.

TABLE I. In-building Carbon Monoxide Model—Parameters and Results

Parameters:	
Condition name: doors part open	
Fans (a-c/hr)	0.5
Doors % open	50
Vent K-factor	2
Building size:	
Average dimension: Width	1000 ft
Length	1500 ft
Height	28 ft
Total Building Volume	42,000,000 ft^3
Blood Carboxyhemoglobin Model	
Time exp. (hrs)	8
Load: Va (ml/min)	18000
DL =	40
Peak COHb (must hit F7 and F8)	28.2
Peterson/Stewart Model (COHb%)	23.4
Coburn Model (COHb%)	27.2
(Coburn includes load and elimination)	
Calculated concentration:	
CO Steady State model	210
CO Build-up model	201
Time (hrs) build-up	8
TWA (build-up model)	177
Climate	
Pressure (mmHg)	760
Temperature (F)	75
Temperature (K)	297
Wind speed (mph)	5

TABLE II. In-building Carbon Monoxide Model—Ventilation

Forced ventilation:	
Number vent. exhaust fans operating (out of 40)	20
Average ft^3/min per fan	16000
Forced exhaust air (ft^3/hr)	19200000
Number of makeup air units working (out of 30)	2
Average capacity of MUAs (ft^3/min)	6000
Percent fresh air (balance is recirculated)	100
Forced supply air (ft^3/hr)	7200000
Larger of exhaust or makeup air (net forced ventilation)	19200000
Air changes per hour (larger of exhaust or supply)	0.46
Natural ventilation:	
Total openings in long side (sq ft)	540.6
Total openings in short side (sq ft)	360.2
Wind direction (degrees) with respect to long side	45
Wind speed (mph)	5
Effective opening (sq ft)	636.7
Natural ventilation (ft^3/hr)	13784357
Total ventilation (ft^3/hr)	33104357

TABLE III. In-building Carbon Monoxide Model—Sources of CO

Source Type	Cycle	Number	CO (g/hr)	Total CO (g)
Concrete breakers, diesel	0.50	4	973	1946
Bulldozers, diesel	0.75	3	660	1485
Tractors with buckets (note)	0.75	8	2647	15882
Backhoes (note)	0.75	10	3656	27420
Hauling trucks, diesel	0.90	12	39	421
Hi-lo's, propane	0.90	12	43	464
Welding machines	0.50	12	5700	34200
Air compressors (note)	0.50	12	2949	17694
Small water pumps, gas	0.90	12	750	8100
Cement trucks, diesel	1.00	4	240	960
Track loader, diesel	0.50	1	251	126
Crane, gasoline	0.40	1	7720	3088
Scissor lifts, gasoline	0.10	24	750	1800
Total CO emissions (g/hr)				113586
Total CO emissions (ft^3/hr)				3477

CHAPTER 18

THE SAFETY AND HEALTH ASPECTS OF VIDEO DISPLAY TERMINALS

SCOTT E. MERKLE, CIH

U.S. Department of Agriculture, Office of Finance and Management, Safety and Health Management Division, Washington, DC

VIDEO DISPLAY TERMINALS

A microcomputer-assisted training program has been developed that focuses on the safety and health concerns of using video display terminals (VDTs). With the advent of computer and information processing technologies, the use of the VDT is becoming increasingly widespread. The VDT, a relatively new tool in many office environments, is an electronic device that conveniently displays alphanumeric and graphic information. The principal applications of these devices are information storage and retrieval in computer systems, and text entry and editing in word processing systems. Today, the VDT is found in nearly all types and phases of business and office work. The National Research Council, National Academy of Sciences has estimated there are currently more than 7 million VDT operators. These estimates are projected to dramatically increase to 40 million VDT users by 1990. Within the Department of Agriculture, plans are under way to introduce "office automation" into 90 percent of Departmental activities over the next three years.

The VDT has presented many office workers with new visual tasks and, in some cases, has brought about significant changes in daily work activities. With the rapid introduction of VDTs, some concerns have been expressed on the impact these devices may have on the operator's health. VDT operators have reported a variety of symptoms including headache, fatigue, eye strain, neck and shoulder pain, and feelings of anxiety or psychological stress. These complaints, along with some widely publicized instances where VDT operators have had newborn children with birth defects, have generated fear among VDT operators. Many operators now do not view the VDT as an office tool that saves time and improves productivity, but rather as a threat to their jobs and their health. The controversy surrounding the VDT, especially reports of radiation hazards, has become an emotionally charged issue. Unfortunately, this has tended to give the VDT user a distorted perspective on the extent of the VDT health risk. We, as safety and health professionals, must educate the VDT operator and dispel some of the myths and fears associated with these devices. The object of operator fears, the VDT, can be used as the agent to dispel those fears through microcomputer-assisted training programs.

MICROCOMPUTER-ASSISTED TRAINING

The microcomputer provides a new and largely untapped method of conveying industrial hygiene related training to employees. The characteristics of the microcomputer learning experience make it a suitable training method in a variety of situations. The key characteristics are listed below:

Nature of information to be conveyed — Programmed instruction techniques are most effective when the training objective is to inform or provide specific information, rather than to influence behavior or provide self insight.

Learner participation — The ability of the learner to participate and make choices will cultivate and maintain interest in the subject matter. Films, slide-tape, video tape, and similar training methods provide little opportunity for the individual to actively participate in the learning experience. In microcomputer-assisted training, some degree of interaction with the program can be achieved by having the user respond to questions and program prompts. However, since the program has predefined ways of responding to user-inputted information, the extent of learner participation is limited, as compared to lecture, laboratory, and simulation training methods.

Pacing — A key disadvantage of conventional training methods is that the pace of learning is controlled by the trainer. Microcomputer-assisted training programs allow the individual to spend as much time as desired on a particular topic, to review a previously covered point, or to select the order in which topics are presented.

Feedback—During the presentation of training sessions, the good instructor often looks for feedback from the participants for an indication whether or not the concepts and information are being understood. Microcomputer-assisted training, along with film, slide-tape and video tape methods, can not make use of this feedback. However, an advantage of microcomputer-assisted training, not shared by other training methods, is the ability to provide immediate feedback to the learner. By incorporating in the program questions to which the user responds, the user can test his/her own understanding of the material.

The computer-assisted learning experience can be designed to be both self-paced and interactive. This flexibility provides a distinct advantage over conventional training methods. However, it may not be the method of choice in all situations. In some cases, it should be supplemented with written manuals or guides and traditional training approaches. During times when safety and industrial hygiene training resources are limited, microcomputer-assisted training offers an economical means of providing training with a consistent message to a geographically dispersed target audience.

PROGRAM DESCRIPTION

Following the title and introductory screens, the user is presented with a menu having eight program selections. Menu entries 1 through 4 cover the key topics of concern to the VDT operator: radiation, effects on vision, physical effects (i.e., ergonomics), and psychological factors. These sections attempt to answer the common questions on VDT health risks. They are intended to be brief and concise, each consisting of 4 to 6 screen "pages" in length.

Menu entry 5 provides recommendations on the design and operation of VDT workstations for comfort and efficiency. In large part, the recommendations reflect those issued by the National Institute for Occupational Safety and Health. This section addresses the major aspects of VDT design and operation: the VDT (equipment), workstation (furniture), work environment, and operator—task considerations. It is 12 screen "pages" in length.

A list of references can be obtained by selecting menu entry 6. The list is intended to serve two purposes; first, to indicate the scientific basis for the information in the program, and second, to assist the user in obtaining more detailed information on VDTs and operator safety and health. If desired, the program will provide a printed copy of the reference list.

Menu entry 7 initiates a question and answer session designed to test the user's understanding of the program material. Ten questions are presented. After the user responds to each question, the program indicates whether or not a correct answer was given. Following an incorrect response, the program can display the screen containing the correct information from one of the previous sections. The program provides a test score after all ten questions have been answered. Operators with scores less than 70 percent are encouraged to review the program.

Menu entry 8 is used to exit the program. The user can return to the menu from anywhere in the program. Function keys are used to display the menu (F7), display the next screen (F10), and display the previous screen (F9).

The program was written using the IBM BASICA interpreter (Advanced Basic language, Version 2.1). The program has been compiled with the microsoft QuickBasic Compiler creating a directly executable file on IBM and true compatible personal computers and requires 116K of memory, in addition to that used by the operating system. The screen displays are in medium resolution enabling the use of basic graphic functions and statements. Medium resolution also provides a large text size which vastly improves the legibility of information on the screen. However, the use of the graphics mode within the program imposes additional hardware requirements: namely, a color monitor and color/graphics adapter card. This limits the number of personal computers that can run the program. Plans are under way to modify the program so that it can be used with a greater number of microcomputers.

CHAPTER 19

A COMPUTERIZED SYSTEM FOR TRACKING FIELD EQUIPMENT MAINTENANCE

MARTIN T. ABELL

National Institute for Occupational Safety and Health, Cincinnati, Ohio

A system using barcodes and a microcomputer data base has been set up to help track the maintenance of IH field equipment. This system will be referred to as CMTS (Computer-based Maintenance Tracking System). The immediate goals of the CMTS are (1) to ensure that the equipment items are checked and reconditioned at regular intervals and, (2) to organize their repair history in computer-readable records. The ultimate goal is to identify weaknesses in either the equipment or the repair schedule, and to improve overall equipment reliability. This, in turn, leads to increased data quality and decreased failure rates, or in other words, improved productivity.

An older system of preventive maintenance has not been replaced, but enhanced. That system included a catalog of written maintenance procedures, a written repair record for each item repaired, and stickers on each item indicating the date of repair and the date of the next repair. The CMTS intrudes on the older system twice corresponding to the two goals stated above. The first is to aid in identifying items for scheduled maintenance and the second is to record the repairs made on each item into a data base. The repair records accumulated in this way can be used to find areas that need improvement and the CMTS provides summary reports for this purpose. Hereafter, the equipment items will be referred to as

Figure 1. Barcode example.

pumps; personal sampling pumps were chosen first because of their high failure rate, which leads to lost data and time.

HARDWARE AND SOFTWARE

The equipment for this project consists of two microcomputers and a barcode reader. One computer is an IBM XT with an HP Thinkjet printer. The data base and barcode printing programs are on this computer. The other computer is an HP 110 Portable. It is programmed in BASIC to communicate with either the barcode reader or the IBM XT. The barcode reader (Digitronics BCR 232) is a wand-type reader attached by a coiled cord to a small box that contains the decoding and transmission circuitry. Transmission is in serial ASCII to the HP 110 RS-232 port. Since the barcode reader and the HP 110 are both battery powered, they can be taken together (usually on a cart) as a portable, intelligent reader. Communication between the IBM XT and the HP 110 is through the HP serial link (HPIL) using a special interface card and software package (HP 82973A) in the XT. When the two computers are connected, the HP 110 can access the IBM XT's disk drives, screen, or printer as if they were attached to the HP 110.

Barcodes

Attached to each pump is a label containing a unique number in barcode form (Figure 1). The barcode is of the "3 of 9" type which allows letters as well as numbers. These barcodes are useful for finding a given pump since they are prominent and can be read with a barcode reader accurately and relatively quickly. The other place where barcodes are used is on a page of repair codes. Common repair procedures are abbreviated to three letters and put in barcode form. For example, replacing the *i*nlet *f*ilter with a *n*ew one is coded IFN (Figure 2), and painting the *ca*se (*r*econditioning) is CSR.

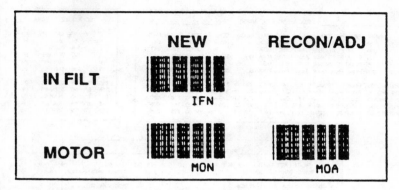

Figure 2. Example of barcode used on a page of repair codes.

Data base

The other element of the CMTS is the computerized data base. The software system being used for data base management is dBase III. It is a command-oriented system rather than menu driven, and the commands are quite powerful. Since the commands can be combined into programs, it is possible to build a complex, customized, data-management application (including customized menus) that is also easy to use. The CMTS consists of half a dozen program files including the one that presents a menu allowing the user to select functions (other programs) to be performed. The data that these programs operate on is actually in three data base files. All three of these "related" files can be open at once to access all the required information (Figure 3). To date, 219 pumps have been entered in the system, and 154 of them have been repaired.

OPERATION

The operation of the CMTS is automated as much as possible using DOS batch files, dBase programs, and BASIC programs that prompt the user at each step.

Identifying equipment

Each month a dBase program called PMLOCATE is run. It searches the Schedule file for pumps whose last recorded preventive maintenance (LAST-PM) was earlier then the present date minus the assigned interval

204 MICROCOMPUTER APPLICATIONS IN OH&S

```
_____DESCRIPTION DATABASE_____
    PROP#  DESCRIPTION              MAN   MODEL  SER#    COST   ROOM AREA
  1     54  PUMP PERSONAL SAMPLING   MSA   G      00054   330.00 108N  047
  2    124  PUMP PERSONAL SAMPLING   MSA   G      00124   330.00 108N  047
  3    127  PUMP PERSONAL SAMPLING   MSA   G      00127   330.00 108N  047
  4    176  PUMP PERSONAL SAMPLING   MSA   G      0176    330.00 108N  047
  5    235  PUMP PERSONAL SAMPLING   MSA   G      00235   330.00 108N  047
  :     :   :                         :    :       :       :      :    :
215  842931 PUMP PERS SAMP, HI FLOW  SKC   224-03 326-36  700.00 108N  047
216  860590 PUMP PERS SAMP, HI FLOW  SKC   224-03 326-35  700.00 116   026
217  860592 PUMP PERS SAMP, HI FLOW  SKC   224-03 326-41  700.00 116   026
218  860594 PUMP PERS SAMP, HI FLOW  SKC   224-03 326-54  700.00 116   026
219  860702 PUMP PERS SAMP, LO FLOW  DUPT  P200   7538    350.00 116   026
```

```
_____SCHEDULE DATABASE_____
    PROP#  LAST_PM   PM_FREQ
  1     54  06/19/85    24
  2    124  06/19/85    24
  3    127  06/19/85    24
  4    176  06/19/85    24
  5    235  11/19/85    24
  :     :     :         :
215  842931 07/19/85    24
216  860590 07/19/85    24
217  860592 01/06/86    24
218  860594 07/01/85    24
219  860702 05/20/85    12
```

```
_____PMHIST DATABASE_____
       PROP#   LAST_PM  R-1 R-2 R-3 R-4 R-5 R-6 R-7 R-8 R-9 OTHER
  1     82910  05/01/85 BAN .  .   .   .   .   .   .   .   FLOW COMPENSATION
  2    427031  05/01/85 BAN .  .   .   .   .   .   .   .   PRESSURE REG ADJ.
  3   8605901  05/01/85 BAN BDN  . .   .   .   .   .   .   PRESSURE REG ADJ.
  4      7128  05/02/85 BAN CJN  . .   .   .   .   .   .
  5      7152  05/02/85 DIN VLN  . .   .   .   .   .   .
  :       :      :       :   :   :   :   :   :   :   :
150     12438  12/06/85 VLN ROR CSR .  .   .   .   .   .
151     82532  12/06/85 ACN VLN IFN SWN CRA BCR .   .   .
152      5089  12/09/85 VLN ROR CSR .  .   .   .   .   .   NEW TUBING
153      5313  12/11/85 VLN CSR ROR .  .   .   .   .   .
154      9028  12/11/85 VLN ACN IFN BCN MON SWR CRA .   .   NEW TUBING
```

Figure 3. Example of three related data base files open at one time.

(PM-FREQ in months). The list of pumps produced is printed and saved on disk. The list on disk can be downloaded quickly to a BASIC program on the HP 110. The HP 110 screen shows the room numbers where the pumps are located. The HP 110 with the barcode reader is taken to those rooms and the barcode on each pump is scanned. When a barcode is read that matches a pump number on the list, the HP 110 beeps and displays the number, otherwise it does nothing. The selected pumps are put on the cart. Counters at the bottom of the screen keep track of the number of pumps found in each room. When as many pumps as can be found are on the cart, they are wheeled to the repair room, and the computers are shut off.

Recording Repairs

After some of the pumps have been repaired and tested, the repairs are recorded as follows. A BASIC program called PMLOG on the HP 110 prompts for a pump number to be read, then for repair codes to be read. Up to 9 three-letter codes (barcodes) and one 20-character field (keyboard) are entered. After entering the repair codes for all the pumps repaired up to that time, the "DONE" barcode is read. Then the HP 110 is connected to the IBM XT via the HPIL link, and the repair data are uploaded. In the IBM XT, a dBase program appends the data to the PMHIST file and updates the SCHEDULE file so that LAST PM is correct for the pumps just repaired.

Reports

Selecting "Report" from the data base menu on the IBM XT produces a summary report of the repairs done to date. It tells, for each type of pump, and overall, how many batteries were replaced, connecting rods adjusted, etc.

CONCLUSIONS

To date, the only consistent problem area identified is one that we were already somewhat aware of, namely, battery performance. A common practice of some field groups has been the measurement of battery terminal voltage to determine battery condition. This voltage reading, particularly under no-load conditions and just after recharging, is a worthless indicator of field readiness. We are presently working on several projects that address this problem and will report on them in the future. We also plan to track battery packs in the CMTS as items separate from the equipment they power.

The CMTS was started a short time ago and is evolving. With regard to pumps, two additions are planned. One is simply to expand the number of pumps in the data base. The other is to add a log-in/log-out system that not only tells when pumps were taken on surveys, but also records any comments about the performance of each pump made by the user. Other kinds of equipment will be included in the CMTS data base, battery packs being the first priority. With the additional data, the possible detection of problem areas should be improved.

Computer Applications Glossary

CHAPTER 20

COMPUTER APPLICATIONS GLOSSARY

KENNETH S. COHEN, PhD, PE, CIH
El Cajon, California

This glossary listing is an abbreviated compilation of terms, critical to dealing with the subject referenced as the above title.

Access — The ability to use a computer or program to store or retrieve information.

Access, random — Information that need not be retrieved in the order in which it was written or stored.

Access, sequential — A storage method (such as on magnetic tape) by which data can only be reached or retrieved by passing through all intermediate locations between the current one and the desired one.

Anova — "Analysis of variance," a technique in which the total variation of a set of data is assumed to be influenced by different causes; the variance due to each cause is separated from the total and measured.

Aspect ratio — The ratio of image height to image width when displayed on an output device.

Backup — Copying of one or more files onto a storage medium for safekeeping should the original become damaged or lost.

BASIC — Beginner's All-purpose Symbolic Instruction Code; a programming language that, due to its ease of learning, is widely taught as a first program tool. It is the principal language used in most microcomputers.

Batch processing—A traditional method of data processing in which transactions are collected and prepared for computer input to process as a single unit. Usually done with a mainframe operation.

Baud—A measurement of communication speeds between devices. It generally means bits transferred per second. Divide the number by 10 to get characters per second (i.e., 300 baud, 1200 baud, etc.)

Baud rate—The number of bits transmitted in one second. A measurement of data transmission speed, expressed in bits per second (bps).

Bit—Either of the characters 0 or 1 in a binary number system; the smallest unit of information the computer can process.

Boot—A shorthand version of "bootstrap"; to enter a routine into the computer's main memory; generally refers to starting the computer.

Box and whisker plot—A graph that displays the range of the central 50% of the data as a box with a line through it representing the median value.

Buffer—Temporary storage for data in a computer's memory. It compensates for differences in speeds of processor and peripheral devices such as storage units, terminals, and printers.

Bug—An error that occurs in a computer program or in the computer's electrical system.

Byte—A group of bits (usually 8). A byte can be used to represent one character (number or letter) of information, all or part of binary numbers, and machine language instructions.

Cathode ray tube (CRT)—A vacuum tube in which electrons are beamed at a fluorescent screen, causing it to glow and thus forming the computer display. It is similar to a television picture tube.

Check bit (digit)—A final digit affixed to a binary sequence for transmission monitoring. It can then be monitored after transmission to ensure the integrity of the received data. *See* parity bit.

Chip—A thin semiconductor wafer on which electronic components are deposited in the form of integrated circuits, and its package of coded signals.

COBOL — Common Business-Oriented Language; a high-level programming language used most often in or on "mainframe" business applications.

Compatibility — The ability of a computer system to accept and process data prepared by another similar system without having to adapt it.

Composite color monitor — A monitor that receives color and synchronization signals mixed and must separate and decode them, losing some picture clarity in the process. Generally composite monitors can display only 40 columns "well."

Computer — A general purpose electrical system designed for the manipulation of information, incorporating a central processing unit (CPU), memory, input/output (I/O) facilities, power supply and cabinet.

Computer-aided instruction — Abbreviated to CAI and is the same as the definition for "computer-based training."

Computer-based training — Abbreviated to CBT; refers to on-screen tutorial instruction on how to use hardware, operating systems and specific applications software programs in a step-by-step, self-paced student-oriented manner.

Configuration — The design of a computer system.

Coordinates, absolute — Coordinates based on points that are x, y, and z distances from a fixed origin.

Coordinates, relative — Coordinates located in terms of x, y, and z distances from an unfixed origin.

CP/M — Control program/microcomputer is a single-user operating system for 8080, Z80, and 8085-based micros; created by Digital Research, Inc.

CPU — Central processing unit; the portion of the computer where the overall activity is controlled.

Crash — A situation where the computer system is made inoperable because of hardware or software malfunction.

CRT — Cathode ray tube; the most popular form of display screen. Similar to a "television" screen display.

Cursor — An electronically generated symbol that appears on the display screen to tell the operator where the next character will appear.

Daisy wheel — A letter-quality printing mechanism whose printing element is made up of a flat metal or plastic wheel with letters molded at the end of its "daisy" spokes or petals.

Data — The basic elements of information that can be processed or produced by a computer.

Data base — A collection of interrelated data organized for ease of update and retrieval. For example, a livestock data base might include the health and breeding information for each animal in the herd.

DBMS — Abbreviation for data base management system. Software that controls storage and retrieval of information in a data base.

Debug — To detect, locate, and correct errors in a computer program.

Default — An option or command assigned by the program or system that is automatically utilized if no other, overriding command is entered by the user.

Density — A term used to describe the distance between magnetic information on tapes or floppy disks. Higher density will increase information storage capability.

Digitize — To input and register coordinate information about an image so that it may be processed by a computer.

Digitizing tablet — A tablet with a cross hair or stylus pointer that is used to input information about an image's coordinates.

Disk drive — A peripheral device that stores information on magnetic disks. Can either be internally mounted or a stand alone unit.

Disk, hard and floppy — A circular plate of magnetic material. This plate rotates for the storage and retrieval of data by one or more "heads" which transfer information to and from the computer. Also known as a diskette or fixed (hard) disk.

DOS — Disk operating system; operating system instructions stored on disk, not in memory (e.g., PC-DOS, APPLEDOS, or MS-DOS used in their respective computers.)

Dot command — A command that uses a period in combination with other letters and keys to give instructions to the computer. Dot commands are used in certain programs such as WordStar®.

Dot matrix — A printer type using a number of pins impacting a ribbon or specially treated paper to form characters.

Dot pitch — The diameter of the dots or picture elements (*see* pixel) that form the monitor display. The smaller the dot, the finer the display. Dot pitch is an indication of resolution or sharpness of the monitor.

Double density — Storing twice as much information on a floppy disk as other standard disk systems.

Double-sided — A floppy disk system that can store data on both sides of a disk, doubling its storage capacity.

Downtime — The period during which a computer is not operating.

Draft quality — Type from a dot-matrix printer with clearly discernible individual dots. Not of the quality represented by an impact printer or typewriter.

Duplex, full — Simultaneous two-way independent transmission between two points, in both directions.

Duplex, half — One-way communication between two points, in either direction.

Editor or/Assembler — A program that manipulates text information and allows the user to make corrections, additions, deletions, and other changes. Sometimes used as an editor/assembler use in assembly level programming.

EIA — The abbreviation for the Electronics Industry Association. Often referred to as a "standard." Address: 2001 Eye Street, NW., Washington, DC 20006.

Encrypted disk — A disk that is programmed in such a way that only authorized users (for instance, with knowledge of the correct entry code or password) can access the disk's files.

Entity — A drawing element, such as a line, circle, or arc, that is listed in a "CAD" data base.

Error message — A statement flashed on the computer screen indicating that the user has done something wrong. Also indicating that the program is unable to perform as instructed.

Exiting — The ability to abandon a procedure if, for example, it becomes too complex or the user has made an error.

File — A collection of related information that is given a specific name and considered a single unit by the computer. It can contain both data and programs.

Floppy disk — A small inexpensive magnetic disk used to store and record information together with a disk drive.

Format — The layout, design, and specifications of a printed document including type size and style, margins, headings, etc. Also a "DOS" command used to set up new disks.

Fortran — FORmula TRANslation. A high-level programming language best suited for complex mathematical applications.

Function keys — Additional keys on a keyboard which are used to perform user or program definable operations.

Graphics — Presenting information pictorially rather than alphanumerically.

Handshaking — Communication between two devices which inform each other about the status of data being transmitted. Used to ensure orderly operation during asynchronous transmissions.

Hard copy — A paper printout. ("Hard" in that you can actually hold it as opposed to copy you can only read on a screen.)

Hardware — The physical apparatus that makes up a computer, including silicon chips, transformers, boards, and wires. The term is also used to describe pieces of equipment like the printer, modem, and CRT.

Histogram — Generally a bar graph, on which the x-axis is divided into ranges of data values. The y-axis shows the frequency or percentage of the total number of samples for each range.

I/O — Abbreviation for data input and output.

Input — (1) The data that is entered into programs. (2) The act of entering data into a computer. (3) Data used by programs and subroutines to produce output.

Input device — Any machine that allows you to enter commands or information into the computer. An input device could be keyboarded, tape drive, disk drive, microphone, light pen, digitizer or other electronic sensor.

Inquiry — A request for information stored in the computer.

Integrated software — One program that contains other programs and permits their simultaneous use or the transfer of data between them.

Interactive — A program or computer that communicates with the user or another computer, i.e., via menus.

Interface (electrical) — Electrical interconnection between system elements.

K — This letter represents the number 1024; abbreviated as "K," "KB" or kilobytes.

Kb — The abbreviation for kilobits, or 1024 bits.

Language — Any set of characters used to perform related commands or instructions that combine into meaningful communications acceptable to a computer.

Layers — Levels of a "CAD" drawing that can be superimposed on a display or plotter.

Least squares — A method of regression in which a formula is derived by minimizing the squares of the differences between the actual data values and the formula approximation.

Letter-quality — A printer output that looks similar to the output of a typewriter.

Load — The process of putting data into the computer or its memory.

Local area network (LAN) — An interconnected communications system that links standalone microcomputers, peripherals and software programs to ease functions such as electronic mail, and data swapping between users in same or other offices.

Log on — To enter the necessary information (a password or other user information/identification number) to begin a session at the terminal or computer.

Mainframe — The largest of computers. With an expansive internal memory and fast processing time. The cost ranges into the millions of dollars.

Matrix — The pattern of dots or pixels used to form individual characters in a display. Most monitors use a matrix of five dots across by seven rows down. Also the format used for characters on a dot printer.

MB — The abbreviation for megabyte.

Memory — The section of the computer where instructions and data are stored. Each item in memory has a unique address that the CPU can use to retrieve information.

Microcomputer — A small but complete microprocessor-based computer system, including CPU, memory, input/output interfaces, and power supply.

Minicomputer — A small computer, intermediate in size between a "micro" and a mainframe computer.

Modem — A device which allows computer equipment to communicate via standard telephone lines.

Modem, acoustic — A type of modem that allows one computer to communicate with another computer or terminal device via telephone. The handset of the telephone is placed into this unit.

Modem, direct connect — In contrast to an acoustic modem, a direct connect modem is hard wired directly to the data transmission line.

Monitor — The computer's screen that displays vivid characters and on which information stored in the computer can be read or viewed.

Mouse — A small, hand-held data input device is rolled on a table top to move a screen cursor and/or enter commands to the computer.

MS-DOS — A disk operating system for microcomputers created by: Microsoft, Inc. MS-DOS is the equivalent of PC-DOS.

Mb — The abbreviation for megabit.

Normal distribution — A distribution of values taking on a symmetric bell-shaped curve centered on a mean that equals the median. The standard normal distribution has a mean of "0" and a standard deviation of "1."

Objects — Drawing elements, such as furniture components, that are stored in a graphic data base and placed in drawings as needed.

On-line — Time spent accessing, retrieving, or storing information on your computer.

Operating system — A collection of programs that controls the overall work of a computer system. (See Disk Operating System — DOS).

Output — Any processed information coming out of a computer via any medium (print, CRT, etc.) or the act of transferring information to these media.

Output device — A machine that transfers programs or information from the computer to some other medium, e.g., tape, disk, printers, typewriters, plotters, VDT, robots, sound synthesizers, etc.

Overhead — The amount of the computer's memory required to organize the operation of the system.

Parallel — In communications, the method of sending an entire character or word at a time over a series of computer lines rather than breaking them up into their component elements. Often believed "faster" than serial.

Parity — Equivalence in the check digit of the transmitted and received data.

Parity bit (check bit) — A binary digit appended to an array of bits to make the sum of the bits always odd (odd parity) or even (even parity). *See* check bit.

Parity checking — Using parity to check for single-bit errors.

Path — An MS-DOS command that tells the disk operating system (DOS) which subdirectories to use (and in that order) when looking for programs that are not in the current subdirectory.

Peripheral — Equipment distinct from the central processing unit of a data processing system which provides the system with additional facilities, e.g., printer, hard disk, etc.

Pixel — Short for picture element, these dots are grouped into rectangular divisions on a video screen. The smaller the rectangle, the more required to form an image, hence a sharper picture.

Pixel, CAD — On a raster (video) display, the smallest resolvable point of a displayed image that can have a specified color or intensity level.

Plug-compatible — Two devices that can work together without a separate interface device.

Portables — Refers to a category of computers that are characterized by their small size and transportability; lab size, briefcase size, and hand-held are all types of portable computers.

Printer — A computer attachment that produces printed copy such as numbers and words on paper.

Program — A sequence of instructions directing a computer to perform a particular function; a statement of an algorithm in a programming language.

Puck — The hand-held, cross-haired pointing device, used with a digitizing tablet to input data.

RAM — Random Access Memory. A type of memory that does not follow a sequence of storage locations to access a location directly vs. sequential access.

Real time — Refers to events as they are actually happening (in comparison to computer time measured in millionths of a second).

Record — A collection of data items stored on a disk or other medium which may be recalled as one unit. Records may be of either fixed or variable length (a data base definition).

Regression — An analysis that expresses a dependent variable as the result of a formula approximation involving one or more independent variables. Linear regression fits a straight line between two variables.

Resolution — The sharpness of the monitor display. It is usually measured by the number of pixels available horizontally and vertically.

Response time — The time required for the system to respond to a user's request or to accept a user's inputs.

RGB monitor — Color monitors that receive red, green, and blue signals from the computer via separate channels. The signals are then combined to form the display colors. RGB displays are usually 80 columns wide.

ROM — Read only memory. A program permanently etched on a computer-integrated circuit chip.

RS-232 interface — The EIA standard for "Serial" Data Transmission. Two key EIA documents are: EIA Standard RS-232, and EIA Bulletin #9.

Rubberbanding — A "CAD" feature that creates a movable line, fixed at one end point, that contracts and expands to follow the screen cursor movements.

Scatter plot — A graph in which the correlation between variables is illustrated by dots (sometimes connected) with a variable along each axis. Also called an "x-y graph."

Serial — The handling of data, one item after another. In communications, a serial transmission breaks each character into its component bits and sends these bits, one at a time to a receiving device reassembling.

Software — A general term for computer programs and documentation involved in the operation of the computer.

Software, application — A program written to accomplish a specific task, such as a payroll, an accident inventory, or an asbestos survey log.

Software, bundled — Software programs that are packaged and sold together with a specific hardware item (e.g., computer + operation DOS, modem + communications software, etc.).

Software, custom — Computer programs prepared for a specific tailor-made purpose. Contrast with packaged software, in which programs are written in advance, usually for general purposes.

Software, integrated — Software that allows the user to move between two or more functions without the need to first save data from the function in use (e.g., Lotus 1-2-3® allowing data base, spreadsheet, and graphic function).

Source code — The humanly readable computer commands written in a program language. It requires an interpreter or compiler. It is sometimes referred to as a source program.

Standard deviation — The square root of the variance.

Storage — The general term for any device which is capable of holding data which will be retrieved later.

Storage, external — Used to store programs and information that would otherwise be lost if the computer were turned off; for example, tapes and disks. Also known as mass storage.

Tape — Inexpensive mass storage medium. Must be accessed sequentially.

Teletext — Textual information transmitted to people's homes via their television. Information is usually maintained and updated on a computer.

Template — Often provided with specific software programs, a template is much like a master worksheet used to format formulas and other basic information for that particular program.

Terminal — A "slave" controller for a computer, usually consisting of a CRT and keyboard. It may be used alone or in combination with several other terminals.

Terminal, intelligent — Smart terminal. A terminal that has some data processing capability or local computing capability, or an actual computer.

Timesharing — A method of sharing the resources of the computer among several users, so that several people can appear to be running different computer tasks simultaneously.

Tractor feed — An attachment used to move paper through a printer. The roller that moves the paper has sprockets on each end that fit into the fanfold paper's matching pattern of holes.

Tree — A DOS operating command that allows the user to see the relationship of the program's directories to each other, as well as a full list of the names of all subdirectories contained in that disk.

UNIX — A computer operating system, developed and owned by AT&T, considered a "powerful manager" of the internal housekeeping chores of multiuser systems.

Upwardly compatible—In reference to software program releases or versions; the capability of most, if not all, operations performed on one version to be performed on the others.

User friendly—Hardware or software with simple instructions that help the user become familiar with the computer.

Utility program—A program that performs a task required by other programs; an example would be using a utility to copy data residing in another program, check disk speed, fix list data, etc.

Variance—A measure of the dispersion of a set of data around the mean, it is calculated by summing the squares of the differences between each value and the mean, then dividing by the number of values.

Video display terminal (VDT)—A CRT plus keyboard.

Window—A software program feature that allows a user to view and, in most cases, manipulate several different files on a terminal screen without having to first store any data.

Word—A unit of data or the set of characters which occupies one storage location. In microcomputing, a character, a word, and a byte are interchangeable. In most minicomputers, a word is equal to two bytes.

Word coordinate system—A coordinate system unrelated to an input or output device.

Word processing—A text-editing program that allows the writing and correcting of text and typesetting.

Word processor, dedicated—A work station word processor whose design and function is solely for word processing tasks.

MICROCOMPUTER INDEX

A
analog to digital converters 44
artificial intelligence 99–100
AutoCad 160

B
BASIC 51
barcode, description of 202
barcode readers 44,202
    Digitronics BCR 232
baud rate 141
bits
    data 141
    stop 141
browse feature 124
    application of 126
buffer 145

C
CAD, see computer aided design
CADKEY 160
carboxyhemoglobin levels 194
    also see COHb
CGA, see color graphics adapter
chemical/material identification 80
chemical/material inventories 81
    organization of 82
    updating information 84
citation tracking, example of 155–158
clock speed 40
CMTS 201
    also see computer-based maintenance tracking system
    dBaseIII 203
    hardware 202
    operation of 203–205
    software 202
COBOL 51
COHb levels 194
    also see carboxyhemoglobin levels
    Ott and Mage model: 194
Coburn equation 194
color graphics
    adapter 40
    high resolution 42
CompuServe 144,146
    On-Line IH Forum 149
    SafetyNet 149
computer aided design 159–163
    applications in IH: 162
    applications in ergonomics: 160
computer communications 133–151
    ASCII file transfer 145
    asynchronous 140
    baud rate 141
    buffer 145
    collaborative authorship 150
    connection 136
    considerations of 47–48
    correction protocols 143
        file level 143
        link level 143
    data word format, definition of 141
    definitions used in 132
    duplex, full/half 141
    equipment, see hardware
    getting on line 141
    handshaking 145
    hardware 134,137
    IH professionals, affect on 147
    interference 136
    media exchange 47
    modems 47
    networks 48
    on-line time 145
    protocol converters 48
    support 50–51
    software 135
    terminal emulation mode 48,142
    transmission errors 143
computer hardware, also see hardware
    mainframe 59

microcomputer
  IBM-PC 60
  IBM-PC/AT 60
  IBM-PC/XT 60
  portable 44
computer software, also see
  software 59
computerized maintenance tracking system 201–205
  goals of 201
conferencing, electronic 148
corrections protocol
  file level 143
    Kermit 144
    Modem-7 144
    Telink 144
    X-modem 143
    Y-modem 144
  link level 143
    MNP 144
    X.pc 144

D

data
  alphanumeric 4
  analogous 72
  destruction of 10
  exposure 109
  handling, see database, spreadsheet
  health 70
    integration of 103–109
  industrial hygiene, integration of 103–109
  numeric 4
  protection of 10
    backup copies 10
    password 11
  record deletion 7
  quality 3
  risk 70
  safety 70
  searching 7
  sorting 8
database
  creation of 4
  industrial hygiene 11–26
  interrelational, use of 165–168
  management 3–26,121–126
  management programs 3,121
  multiple 76–77
  respirator, creation of 122–126
  toxic substances, creation of 166–168
data entry
  considerations 43–44
  error checking schemes 6
  fields 4
  modification 7
  procedures of 4,6
    coding 4
  record format 4
  records, see fields
data management, also see
  data base environment 69
data terminal equipment, also see
  hardware 137
data/computation path 40
DCE, see data communication equipment
DDS, see decision support system
DEChealth 98
decision support systems 34–35
digitizers 44
display feature 124
  application of 126
DTE, see data terminal equipment
download 145
duplex, full/half 141

E

EDP, see electronic data processing
electronic conferencing 148
electronic data processing 32–33
electronic mail 148
electronic publishing 150
environment, definition of 72
Environmental Protection Agency 95
exposure, acute/chronic 73
employee, history of 104
  monitoring of carbon monoxide 193–196
    build-up model 194
    steady-state model 194

F

field 4,122
file 122
FLOW GEMINI 98–101

INDEX

Fortran 51

G
Generic CADD 160
graphics output, considerations of 47

H
handshaking 145
hardware
    mainframe 39,95
    microcomputer 38
        CP/M 130
        HP 110 202
        IBM-PC/compatible 130,166,175,177
        IBM-XT 202
        Macintosh 177
    minicomputer 38,97-99
    modem 47
    printer
        daisy wheel 45
        dot matrix 45,145
        dot matrix, Apple 177
        dot matrix, Epson 177
        dot matrix, Thinkjet 202
        electrostatic 46
        ink-jet 46
        thermal 46
hardware selection guidelines 48-49
Hazard Communication Rule 93,100
health hazards, known 103
    management of 104
health surveillance 103-109
high resolution
    color graphics 42
    software support 43
    monochrome graphics 42

I
indoor construction
    exposure to carbon monoxide 193-196
information management systems
    environmental considerations 71-73
    right-to-know 71
information processing, defining problems 31-32

information systems
    building blocks 27-57
        hardware 28,38-49
        procedures 29-30,54-56
        software 28
    concept of 30-31
    operator skill level 52-54
    procedures
        data security considerations 55-56
        documentation considerations 54-55
        general guidelines 56
        training considerations 55
    requirements for 73-76
    software 49-52
        data base management systems 51
        electronic spreadsheets 51
        operating system 49
        programming language 51
information, dependable 69-77
input/output 40

K
keyboard 43
knowledge (expert) systems 35-36

L
language, programming
    Advanced Basic, Version 2.1 200
    BASIC 51
    COBOL 51
    compiler 51
    dBase 43,123
    Fortran 51
    IBM BASICA 200
    Pascal 51
Lotus 1-2-3, use of 62-64,155-158

M
macro 146
macro cosmos, definition of 72
mail, electronic 148
mail-merge 126
management information systems 34
mass storage considerations 45
    floppy disk 45
    hard disk 45
    the future 45

material safety data sheets 71
  computerization 85
  management of 84
materials inventory, automation of 79–87
math coprocessor option 40
measurement
  metabolic energy expenditure model 169–175
  static strength model 175–177
metabolic energy expenditure, prediction of 169–175
micro cosmos, definition of 72
microcomputer
  assisted training 198–199
  compatibility 39
  peripheral devices 40–44
microprocessor 39
  Intel 39
Microsoft-Disk Operating System 165
MIS, see management information systems
modem, also see hardware 137–141
  acoustical couple 138
  definition of 138
  direct connect 138
  frequency 139
  modulation, standards of 139
  settings 140
  smart/dumb 138
  use of 137
monitors, also see VDTs
  monochrome 41
    high resolution 42
  monochrome graphics, high resolution 42
mouse 43
MS-DOS, see Microsoft-Disk Operating system
MSDS, also see managerial safety data sheets
  information, automation of 79–87

N
National Academy of Sciences 97
National Institute for Occupational Safety and Health 95
National Research Council 97

O
occupational health surveillance systems
  artificial intelligence, application of 99
  background 92–94
  DEChealth 98
  FLOW GEMINI 95–98
  environmental applications 99
  evolution of 91–102
  framework, historical 94
  implementation of 92
  SENTRY 104–109
  TOHMS 98
  use of 92
office automation 36–38
OHSs, see occupational health exposure systems
OSHA, see Occupational Safety and Health Administration
Occupational Safety and Health
  Act of 1970 95
  Administration 95
on-line data base
  Dialog 147
  IH forum 149
  Medline 147
on-line time 146
operating speed, parameters of 40

P
parity 141
Pascal 51
password, protection of 142
Peterson-Stewart equation 194
pixel, definition of 40
printer considerations 45–47
program model
  carbon monoxide levels 193–196
  metabolic energy expenditure prediction 169–175
  static strength prediction 175–177
programming language, also see language 51
programs, see software
prospective health management 106
publishing, electronic 150

INDEX

R
random access memory, capacity of 29
record, also see fields 122
report 122
resolution 40
right-to-know 100
risk
  health, unknown 107
  known, management of 104

S
scripts 146
SENTRY 104-109
SafetyNet 149
software
  commercial, CAD 159-163
    AutoCad 160
    CADKEY 160
    Generic CADD 160
  commercial, OHSs
    DEChealth 98
    FLOW GEMINI 98,98
    TOHMS 98
  commercial, communications
    ASCII Express 135
    ASCII Pro 135
    ASCOM IV 135
    Apple ACCESS 135
    CrossTalk 135
    CrossTalk-Fast Ver. 3.51 139
    CrossTalk Mark IV 135
    CrossTalk XVI 68,135
    Hayes Smartcom I 135
    MEX (PD) 135
    MEX-PC 135
    MITE 139
    Macterminal 139
    Microsoft ACCESS 135
    Modem 7 (PD) 135
    ProYam 135
    Red Ryder 135
    Relay Gold 135
  commercial, data base management
    dBaseII 165
    dBaseIII 65,122-126,203
    dBaseIII Plus 65
    R:5000 65

commercial, graphics
  Diagraph 68
  Grafix 68
  Sign Master 68
commercial, operating system
  Concurrent PC-DOS 49
  Microsoft basic 177
  MS-DOS 60,66,165
  MS-DOS 2.0 177
  PC-DOS 49,60,66
  Pick 49
  Unix/Xenix 49
commercial, spreadsheet
  Lotus 1-2-3 62,155-158
  Symphony193
  VisiCalc 114
commercial, word processing
  Microsoft Word 66
  Multimate 66
  WordPerfect 66
  WordStar 66,128
communications 68
design of 79-87
documentation consideration 54-55
electronic spreadsheets 51
freeware 62
graphics 67-68
operating system 49
selection guidelines 52
shareware 62,131
shareware, communications
  PC-TALK 68
  PC-TALK III 135
  PC-TALK III-B 139
  PIBTERM 135
  PIBTERM 3.2.5 136
  Q-Modem 3.0 136
shareware, data base management
  PC-FILE III 65
shareware, graphics
  PC-Picture Graphics 67
shareware, word processing
  PC-Write 67,131
  Textra 67
spreadsheet, electronic
  also see software

applications of
  61,113–119,155–158
  air sampling 114
  citation tracking application
    155–158
  industrial hygiene
    management 113–119
    Lotus 1-2-3 62–64,155–158
    noise sampling 117
    Symphony 193–196
    ventilation measurement 118
  vertical 75–76
  word processing 66–67
software support 44
special interest groups, on-line
  On-Line IH Forum 149
  SafetyNet 149
static strength, prediction of
  175–177
steady-state model, carbon
    monoxide exposure 194
system implementation 56–68

T
terminal emulation mode 142
threshold limit value 194
TOHMS 98
touchpads 43
touchscreen 43
TLV, see threshold limit value
transmission errors, correction of
  143

U
upload upload/download 145–146
  direct link 136
  modem transmission 136

V
ventilation forced 196
  indoor 193–196
  natural 196
video display terminals, also see
    CRT, VDTs
  applications of 197
  safety and health, aspects of
    197–200
voice 43

W
word processing 127,197
  applications in IH 127–132
  features of 131
  non-resident programs
    Check-disk 130
    PC-Read 130
    grammar/punctuation 130
  resident programs 129
    pop-notebook 130
    pop-up calculator 129
    print buffer 130
    print spooler 130
    printer controller 129
  text editing 128